SCIENCE, REALITY, AND LANGUAGE

SUNY Series in Philosophy
George R. Lucas, Jr., Editor

SCIENCE, REALITY, AND LANGUAGE

Michele Marsonet

STATE UNIVERSITY
OF NEW YORK
PRESS

Published by
State University of New York Press, Albany

Production by Susan Geraghty
Marketing by Bernadette LaManna

Printed in the United States of America

For information, address State University of New York Press,
State University Plaza, Albany, N.Y. 12246

Library of Congress Cataloging-in-Publication Data

Marsonet, Michele.
Science, reality, and language / Michele Marsonet.
p. cm. — (SUNY series in philosophy)
Includes bibliographical references and index.
ISBN 0–7914–2475–8 (alk. paper). — ISBN 0–7914–2476–6 (pbk. : alk. paper)
1. Science—Philosophy. 2. Analysis (Philosophy) 3. Language and languages—Philosophy. 4. Realism. 5. Naturalism. I. Title.
II. Series.
Q175.M374 1995
501—dc20 94–22357
CIP

10 9 8 7 6 5 4 3 2 1

To Davide and Carlo

CONTENTS

PREFACE

This book was written, with the exception of the first chapter and some sections of chapters 3 and 5, while I was a Visiting Fellow at the Center for Philosophy of Science of the University of Pittsburgh in the spring term of 1993. It is a proof, I hope, of the lively and extremely stimulating atmosphere of the Center, which allows scholars from all over the world to pursue their own research freely and gives them the opportunity to discuss the topics of their interest with some of the major representatives of contemporary American philosophy. In Pittsburgh, where there is still the exciting philosophical climate I found when I was a graduate student in the Philosophy Department, one has constantly the impression of being at the center of the debates currently thriving in philosophy of science, logic, epistemology, and metaphysics.

The central chapters are modified and enlarged versions of two articles that appeared prior to the publication of this monograph: "Is Philosophy of Language Really Important for the Foundation of Scientific Realism?" (*American Philosophical Quarterly*), and "Realism and Anti-Realism: An Old/New Debate" (*Idealistic Studies*). They express well my views about the influence—which I deem negative—of linguistic analysis on the philosophy of science, and on the current debate beween realists and anti-realists in philosophy of science and metaphysics. I wish to thank the editors of the two journals for permitting the partial reproduction of the aforementioned articles. In writing the book I made an extensive use of quotations. I think, in fact, that instead of giving summaries which may sometimes be boring, it is better to let the quoted authors speak with their own words. This method may be criticizable to some extent (someone, for instance, might argue that the quotations interrupt the fluency of the text); still, I believe this is the best strategy for assuring a lively representation of the philosophical discussions currently under way.

I am much indebted, first of all, to Nicholas Rescher who, although he does not agree with some of my theses, spent a considerable amount of time reading the various chapters of this monograph and discussing them with me. His suggestions also prompted me to take into serious account the philosophical contributions of American pragmatism, and now I believe that many issues that are currently discussed within philosophy of science, philosophy of language, and the theory of knowledge may find in that context interesting solutions. And my debt is extended to other members of the Center for Philosophy of Science—especially Adolf Grünbaum, Jerry Massey, and Wesley Salmon—who helped me in my research. With all of the Fellows visiting the Center in the spring term of 1993 I had interesting discussions. I would like to mention Arto Siitonen, who pointed out to me some features of Hans Reichenbach's thought that I had previously neglected, and Malcom Macmillan, who patiently read some chapters of the book and gave me precious linguistic suggestions. The staff of the Center provided me with the ideal atmosphere for working. To my sons Davide and Carlo, and especially to Carlo, who is now 21 months "old" and always smiles, the book is dedicated.

CHAPTER 1

The Analytic Legacy

1. INTRODUCTION

This essay is based on the conviction that some well-known ideas contained in W. V. O. Quine's works played—and are still playing—an important role in the anti-realist stance currently flourishing in the philosophy of science. I am referring to Quine's entire work, although the doctrines of (a) the inscrutability of reference of terms and (b) the indeterminacy of translation of sentences seem to me particularly important in this respect. I know, of course, that other authors have said many valuable things in this regard, but none of them has been as influential as Quine. This happened because (i) his thought maintained for several decades a sort of supremacy in the field of analytic tradition following the crisis of logical empiricism, and (ii) his fascinating challenge to the analytic/synthetic distinction convinced many people that he succeeded in definitely overcoming the strictures of logical empiricism itself. I think, instead, that Quine is still very close to neopositivism, even though this conviction of mine runs into conflict with arguments advanced by other scholars such as, for instance, George D. Romanos.[1]

The fact of the matter is that we cannot even hope to understand the influence of Quine's theses on current scientific anti-realism if we do not get clear on the way he conceives the relations between *language* on the one side and the *world* (reality) on the other. This means that we must explore the philosophical roots of Quine's thought first; only when this exploration is satisfactorily accomplished will we be able to examine the consequences of Quine's ideas on many contemporary philosophers of science (such as Putnam, to take just one example). Furthermore, I would like to clarify another important issue: the fact that Quine's influence on current linguistic anti-realism may be argued for *despite* his official empiricist and proscientific stance.

I know perfectly well that such a strategy is not fashionable nowadays, since it implies getting immediately involved in ontological and metaphysical questions. And the antimetaphyisical prejudices so stubbornly defended by logical empiricism are still at work, despite the novelties brought in by Popper and his followers (and—this must be honestly recognized—by Quine himself). But a tour through the unpopular territories of ontology and metaphysics is absolutely necessary because it would otherwise be impossible to understand what I mean to do in this work. It should be clear from the beginning, however, that this is not a monograph *on* Quine, although his theses are discussed at length, particularly in the second chapter. My purpose is wider and twofold. I would like to show, in the first place, that analytic philosophy of language has had an impact on contemporary philosophy of science which is both enormous and negative. And, second, my intent is to point out that philosophy of science, in the last few decades, has grown as a largely autonomous and purely speculative discipline, one which does not take into account—with a few exceptions—what practicing scientists are actually doing. This explains why physicists like Steven Weinberg and Stephen Hawking are extensively quoted in the book. They both point out, in fact, that philosophy of science as it is currently practiced is of little or no interest to scientists, and I believe that they have extremely good reasons for making such a claim.

In this sense, I agree totally with Ian Hacking when he explicitly underlines that the theories of meaning may be, and have actually been, dangerous for philosophy of science. He claims, in fact, that

> Anti-metaphysical prejudices and a verification theory of meaning are linked largely by historical accident. Certainly Comte was a great anti-metaphysician with no interest in the study of "meanings". Equally in our day van Fraassen is as opposed to metaphysics. He is of my opinion that, whatever be the interest in the philosophy of language, it has very little value for understanding science. . . . Among the distinctive traits of logical positivism is an emphasis on logic, meaning, and the analysis of language. These interests are foreign to the original positivists. Indeed for the philosophy of science I prefer the old positivism just because it is not obsessed by a theory of meaning. . . . Unlike the logical positivists Popper thought that the theory of meaning is a disaster for the philosophy of science.[2]

Starting from these premises, and taking into account what I said earlier, I think it is possible to verify that most contemporary analytic philosophers have come to endorse a sort of "linguistic idealism" which is just the opposite of the empiricist faith they openly proclaim. In my opinion, this widespread trend in contemporary analytic philosophy has clear Kantian roots, and, in fact, the rejection of some of Kant's basic tenets plays a big role in this work. I also discuss at length Nicholas Rescher's pragmatic and conceptual idealism (trying to show that his idealist epistemology is balanced by a clean realist position at the ontological level) and Richard Rorty's interpretation of the second Wittgenstein.

The conclusion of the book is that philosophy of science, in order to rebuild the bridges with operative science that have been almost completely destroyed by the prevalence of the linguistic turn, must turn to naturalism and take into account the vision of scientific activity that practicing scientists hold. It follows then that philosophers of science, if they want to regain the respect of the scientific community, should reject both the overevaluation of linguistic analysis and formal logic maintained by logical positivism, and the purely historical and sociological approach adopted by many representatives of the post-empiricist turn.

2. THE REAL MEANING OF THE LINGUISTIC TURN

The linguistic turn that has so largely dominated analytic philosophy in the twentieth century can be seen as a bold attempt at preserving both the utility and the significance of philosophy conceived of as an autonomous enterprise. Following the "death of metaphysics" as envisioned by the logical positivists, philosophers had a vital need to preserve a field (and a methodology) of inquiry that they could feel was their own, since the Vienna Circle's program left them—or, at least, this is the standard interpretation of the logical positivists' theses—with almost nothing to do. Obviously, what we just said is true only if logical positivism is taken seriously, which happened only to a certain extent within contemporary philosophy considered as a whole. As a matter of fact, most metaphysicians did not even bother to read what the members of the Vienna Circle—along with their German and Polish allies—claimed. Certainly neither Martin Heidegger changed his philosophical outlook after Rudolf Carnap published his famed paper on the "elimination of metaphysics through the logical analysis of language,"[3] nor did Jean-Paul Sartre give up the construction of his ontology due to the alleged lack of meaning of such an enterprise according to the neopositivists.

In my opinion, however, it is a fact that logical positivism addressed a *real* problem, no matter how many of its solutions—if any at all—we are ready to endorse. And this problem is that philosophy has been clearly displaced in most sectors of knowledge by modern sciences, be they natural, historical, or social. This situation is certainly well known, but it is worth noting that the reflection on its real consequences has been thus far largely inadequate. What I want to recall, in short, is that in the model of philosophical inquiry envisioned by the logical empiricists, analysis of scientific language becomes something similar to a metaphysical endeavor which is meant to establish the *bounds of sense*, and this stance may be easily traced back to Wittgenstein's *Tractatus Logico-Philosophicus*. On the other hand, the analytic tradition transferred this conception to the analysis of ordinary language, and this move, eventually, was able to restore the confidence of (at least some) philosophers in their own work. After all, they were doing something important and worthwhile (i.e., something no one else was doing, since linguists are certainly concerned with language, but from quite a different point of view).

At this point we may well ask ourselves: What is wrong with this kind of approach? At first sight it looks perfectly legitimate and, moreover, it produced wonderful results, as anybody can verify just by reading the masterpieces of contemporary analytic philosophy. To answer the question *What is wrong?* we must first of all take into account language itself and check what it is meant to be within the analytic tradition. This will give our question a clear answer. We have to verify, furthermore, what kind of knowledge philosophy needs to be equipped with if it wants to preserve its autonomy. The logical positivists clearly claimed in their program that there is no synthetic a priori knowledge such as the one envisioned by Immanuel Kant. There is, however, an analytic and a priori knowledge which is supplied by mathematics and logic alone. Within this field, the techniques of contemporary formal logic—as envisioned by Gottlob Frege, Bertrand Russell, Alfred N. Whitehead, and many others—are exalted because they allow us to build artificial languages which—at least theoretically—eliminate the ambiguities of everyday speech.

Richard Rorty claimed recently that the linguistic turn was an attempt to find a substitute for Kant's transcendental standpoint.[4] He is right in this respect, but I also think that it would be better to speak of a "natural continuation" of the Kantian viewpoint (in any event, we will examine Rorty's position in detail later on).[5] The question to be asked now is the following: What are the philosophers supposed to do, if the aforementioned vision endorsed by logical posivism is correct? It is clear, in fact, that we cannot invent an activity just to allow philosophers to have a job and survive: this is, after all, a matter of intellectual honesty. In other words, *if* logical positivists and analytic thinkers are right (i.e., *if* philosophy has been completely displaced by science so that it no longer is a kind of knowledge independent of the scientific one), *then* we are bound to conclude that there is no need for philosophy any more. Philosophy departments may be shut down and perhaps replaced by other structures, in which the analysts of language (no longer called "philosophers") would work with scientists, both natural and social. But the previous question is still open: *What* are these analysts supposed to do, after all? Professional scientists do not seem favorable to this project and, as a matter of fact, some of them oppose it openly.

First of all, we must note that scientists themselves are perfectly aware that philosophical issues continue to arise within science, and to such an extent that it is somehow justified to speak of a "rebirth of metaphysics *within* science." Obviously we are not talking, in this context, of classical metaphysics such as the Aristotelian or Thomistic ones, because in our days the figure of the metaphysician who deems it possible to discover the "first principles of Being" just sitting at his desk is definitively outmoded (despite its persistent popularity in many philosophical circles of continental Europe). We mean, instead, a metaphysics (and an ontology) *contiguous to* science to the extent that, when philosophers are not willing to cope with metaphysical issues, professional scientists invite them to change their attitude because they somehow feel that philosophers are the only persons entitled to carry out this job.[6]

This situation is not totally new. If we consider the classical positivism of the past century, it is easy to verify that *mechanism* was a sort of new metaphysics—stemming from natural science—

which was enormously successful not only with practicing scientists but also with many scientifically oriented philosophers. Mechanics, in fact, offered to the positivists the opportunity to build up a unified synthesis of scientific knowledge taken as a whole, thus pursuing the project of explaining any natural fact by means of the mechanistic model. But this, of course, *was* metaphysics, since the positivists thought that they were able to reach the first principles of a reality conceived of in purely material and observable terms. Since reality is formed only by matter—the positivists claimed—science is able to get a complete knowledge of it. So we had a metaphysics which was both *unconscious* and *monistic*, and this fact is likely to teach us something precious about the practical impossibility of eliminating metaphysics once and for all. It is also worth noting that Marxism itself is rather akin to positivism: to the vision according to which the only reality is the natural one, Marxism adds that the character of social reality is uniquely economic.

We all know that logical positivism is just an updated version of classical positivism. The positivists of our century no longer view philosophy as the elaboration of metaphysical world-visions but, rather, as a technical and linguistic activity meant to clarify the meaning of concepts: a pivotal role is played, in it, by formal logic. In this sense, neopositivism is very close to an even wider philosophical movement called analytic philosophy and, as a matter of fact, authors like Rudolf Carnap and Alfred J. Ayer are both logical positivists and analytic thinkers. The two philosophical trends just mentioned have many common characteristics, because both make of linguistic analysis the *only* task pertaining to philosophy. In one case (neopositivism) scientific language is mostly taken into account, while in the other (analytic philosophy) everyday language is the main object of inquiry. In this work we do not mean to explore the strict relationships between these two trends of thought and their common founding fathers such as Gottlob Frege and Ludwig Wittgenstein (to mention just two names out of many). Let us only note that logical positivism is a very important subset of a large set of tendencies which, taken together, form twentieth-century analytic philosophy. Here, we are mainly interested in clarifying the metaphysical tenets of contemporary linguistic analysis (and of logical positivism in the first place), a theme that we shall examine in the next section.

3. WHAT IS PHILOSOPHY FOR LOGICAL POSITIVISM?

According to the neopositivists, the sole true knowledge is empirical and based on immediate observative data; furthermore, they reject the Kantian synthetic a priori, even though, as we shall see later on, Kant's influence on their philosophical outlook is quite strong. As I hinted previously, they attribute a pivotal role to formal logic because, in their opinion, it allows us to formalize in a rigorous manner the intuitive inferential processes of ordinary language. In our day logical positivism is less popular than it used to be until a few decades ago, although it maintains a considerable influence (especially in Great Britain and North America). The so-called post-empiricist turn questioned practically every single point of its general outlook on philosophy and the world (an outlook that is often defined as the *received view*, just to stress the fact that, despite its many shortcomings, it is the starting point of a trend of thought whose importance within contemporary philosophy cannot be denied). Today a successful philosopher of science like Paul K. Feyerabend holds views that are practically opposed to those endorsed by logical positivism. This means, however, that if we want to understand Feyerabend's current popularity, we are bound to read the neopositivists' works very carefully. In my view, it is not correct to claim—as many contemporary authors do—that the neopositivists are *completely* wrong. This is clearly an overstatement, because the members of the Vienna Circle—along with their allies of the Berlin Circle and of the Lvov-Warsaw School—can at least be credited with one great merit: they compelled philosophers to take science seriously into account in a period when it was largely believed that philosophy and science are totally independent fields of inquiry (a position that, unfortunately, many philosophers still endorse nowadays).

If we try to identify the position of logical positivism in the map of contemporary philosophy, we will soon find out that it can be characterized by a few basic and radical theses:

1. Neopositivism is not a philosophical system but, rather, a general *attitude* towards philosophy which denies any validity to the way philosophical work has been carried out in the past centuries.

2. The logical positivists think that philosophy is not a *speculative* discipline; it is, rather, a logico-linguistic activity aimed at clarifying scientific propositions.

3. There are only analytic (a priori) and synthetic (a posteriori) propositions. The first class is formed by logical and mathematical sentences, and the second by the sentences that can be found in the empirical sciences (where physics has a predominant role). As we said before, there is no Kantian "synthetic a priori."

It follows that the whole of human knowledge can be reduced to the two classes of sentences just mentioned, and this means that the only possible knowledge is given by science. Metaphysics is thus meaningless, because its sentences do not comply with the rules set forth by logical analysis of language. What, then, is the philosopher's job? The members of the Vienna Circle answer that his task is to clarify the concepts used within empirical and formal sciences, while analytic philosophers stress instead the importance of ordinary language's analysis. But the outcome is in both cases clear: philosophy *is* linguistic analysis. It may be observed that neopositivism certainly has some ancestors in the history of philosophy: the sophists of ancient Greece, such as Protagoras; the nominalists of the Middle Ages, such as Ockham; the classical British empiricists (and especially Hume); and the positivists of the nineteenth century, such as Comte. Their radicalism, however, is rather new. The logical positivists want to rebuild philosophy *ab initio*, just making a tabula rasa of what has been said and done in many centuries of philosophical speculation. And their attitude is based on two undeniable facts: (a) the enormous results, both speculative (knowledge of empirical reality) and practical (technological applications) accomplished by modern science from Galileo on; and (b) the spectacular achievements of formal logic which, starting from Frege and Russell, set forth the project of accomplishing the Leibnizian dream of the *calculemus*, that is, the complete formalization and mechanization of human reasoning.

The logical positivists of our century are, then, despite their official lay spirit, the prophets of the "new scientific world-perspective." Their stance is a full-fledged *scientism*, and, needless to say, Feyerabend's position (i. e., an antiscientific outlook growing within philosophy of science itself) can perhaps be better understood if we take it to be a reaction to the "received view" of logical positivism.

If we now examine the neopositivists' alleged destruction of metaphysics, it is possible to note that their attack actually missed the target due to its essential vagueness. *Whom*, or *what*, did they mean to attack? It is quite evident, in fact, that the word *metaphysics* has an incredibly high number of semantic and historical connotations. Plato and Aristotle are both metaphysicians, but is this sufficient to associate them? Hegel and Bergson, too, are metaphysicians, but who dares to claim that this fact makes them similar? We must recall that neopositivism is, first of all, a reaction to the predominance of idealism in the Austrian and German academic circles during the last decades of the past century: the real targets of the members of the Vienna and Berlin Circles (and of their Polish allies) are in fact Hegel and their contemporary Martin Heidegger. Taking again into account the above-mentioned essay by Rudolf Carnap, "The Elimination of Metaphysics Through Logical Analysis of Language,"[7] it is easy to verify that, in order to show the purported meaningless of metaphysics, Carnap just analyses from a logical viewpoint some statements made by Hegel and Heidegger:

> Just like the examined examples "principle" and "God," most of the other specifically metaphysical terms are devoid of meaning, e.g., "the Idea," "the Absolute," "the Unconditioned," "the Infinite," "the being of being," "non-being," "thing in itself," "absolute spirit," "objective spirit," "essence," "being-in-itself," "being-in-and-for-itself," "emanation," "manifestation," "articulation," "the Ego," "the non-Ego," etc. These expressions are in the same boat with "teavy," our previously fabricated example. The metaphysician tells us that empirical truth-conditions cannot be specified; if he adds that nevertheless he "means" something, we know that this is merely an allusion to associated images and feelings which, however, do not bestow a meaning on the word. The alleged statements of metaphysics which contain such words have no sense, assert nothing, are mere pseudo-statements.[8]

But it is not difficult to understand that to criticize two particular philosophers does not imply attacking metaphysics as such: if it may be justified to claim that Heidegger often plays with the "magic of words," certainly this charge cannot be addressed to such rigorous philosophers as Aristotle or Leibniz. The situation becomes even clearer if one takes into account the classical essay *Language, Truth and Logic*, written in the 1930s by Sir Alfred J.

Ayer. In this work, in fact, some remarks may be found that are very important for our purposes:

> The belief that it is the business of the philosopher to search for first principles is bound up with the familiar conception of philosophy as the study of reality as a whole. And this conception is one which is difficult to criticize, because it is so vague. If it is taken to imply, as it sometimes is, that the philosopher somehow projects himself outside the world, and takes a bird's-eye view of it, then it is plainly a metaphysical conception. And it is also metaphysical to assert, as some do, that "reality as a whole" is somehow generically different from the reality which is investigated piecemeal by the special sciences. But if the assertion that philosophy studies reality as a whole is understood to imply merely that the philosopher is equally concerned with the content of every science, then we may accept it, not indeed as an adequate definition of philosophy, but as a truth about it. For we shall find, when we come to discuss the relationship of philosophy to science, that it is not, in principle, related to any one science more closely than to any other. In saying that philosophy is concerned with each of the sciences. . . we mean also to rule out the supposition that philosophy can be ranged alongside the existing sciences, as a special department of speculative knowledge. Those who make this supposition cherish the belief that there are some things in the world which are possible objects of speculative knowledge and yet lie beyond the scope of empirical science. But this belief is a delusion. There is no field of experience which cannot, in principle, be brought under some form of scientific law, and no type of speculative knowledge about the world which it is, in principle, beyond the power of science to give. . . . With this we complete the overthrow of speculative philosophy. We are now in a position to see that the function of philosophy is wholly critical.[9]

On the one hand Ayer's statements are very clear, but on the other they make us understand why the elimination of metaphysics could not be carried out (and this also justifies the length of our quotation). Let us take into account, for instance, the concept of "reality as a whole." Ayer remarks that, in claiming to study reality as a whole, the metaphysician pretends to project himself outside the world taking a bird's-eye view of it. Assuming that any serious metaphysician really means to do this (which is, at least, questionable), there is a sentence that clearly reveals Ayer's hidden thoughts. In fact, he goes on to claim that "there is no field of experience which cannot, in principle, be brought under some form of scientific law, and no type of speculative knowledge about the world which it is, in principle, beyond the power of science to give." It is possible to note, then, that in pronouncing these statements Ayer is not talking about a reality which is investigated "piecemeal" by the special sciences. He speaks, instead, of (1) a reality *as such* which has an exclusively empirical character, and (2) a purported unified method which natural science uses in order to investigate reality.

But, at this point, two facts are neatly exhibited: (3) by reducing the whole of reality to empirical reality, Ayer is doing metaphysics. A metaphysical system need not be idealistic: there is an empiricist metaphysics, as is shown by the developments of the last century's mechanism; and (4) even Ayer's reality turns out to be "reality as a whole." Science, as conceived of by Ayer, is in fact nothing but a tool for knowing reality as such. We do not have the "elimination" of metaphysics here, but just the proposal of an empiricist brand of metaphysics. And this fact confirms, once more, that we must distinguish what the logical positivists *say* from what they actually *do*.

4. THE METAPHYSICAL TENETS OF LOGICAL POSITIVISM

Are we allowed to say, however, that the neopositivist program aimed at the elimination of metaphysics has failed? The answer to such a question is not as easy as it might seem at first. From a certain point of view the answer is, "Yes, the program failed and the members of the Vienna Circle did not succeed in getting rid of metaphysics." This is due, as we noted previously, both to the vagueness of their attack and to the fact that a large number of philosophers maintained their interest in this old sector of philosophy no matter what the logical positivists had to say about it. On the other hand, however, someone might answer, "No, the program did not fail completely, because the neopositivists gave us precious insights into the *epistemological* status of metaphysics, so that no serious scholar can ignore, nowadays, their theses on this subject." Who is right? In my view both answers are correct to some extent, and at the same time they are both partial. Let me explain why.

First of all, it is obvious that metaphysics has *not* been eliminated: this is shown by the fact that it is still thriving, even within the analytic tradition itself (as we shall see later on). In addition, as we have just pointed out, it must be clearly distinguished between what the logical positivists claimed and what they really did. It may be argued that, like the classical positivists in the past century, contemporary neopositivists built *their own* metaphysics, which is, as a matter of fact, beautiful and fascinating. If we forget this situation—as, unfortunately, many analytic authors today do—we are not likely to understand the crisis of logical positivism and the subsequent rise of post-empiricism, whose success is also due to the reaction to the particular brand of metaphysics endorsed by logical positivism. Secondly, the fact that the traditional way of doing metaphysics—although it is still practiced in several philosophical circles—seems so outdated nowadays is largely a consequence of the methodological and epistemological criticisms set forth by neopositivists themselves and subsequently endorsed by the whole analytic tradition. Probably it is not correct to speak—as many authors do today—of the "end of metaphysics," but no doubt the new approaches to this discipline, including those currently popular within the analytic school, are much indebted to the neopositivists' insights.

Just to explain what the metaphysical tenets of logical positivism are, we may start with a well-known critique addressed to the so-called principle of verification, which is the keystone of neopositivist methodology. According to this principle a sentence—except for the analytical sentences of logic and mathematics—is meaningful from the cognitive point of view if, and only if, its truth or falsity may be established by having recourse to empirical observations. This is the basis of the neopositivist elimination of metaphysics, since according to the representatives of the Vienna Circle metaphysical propositions, along with ethical and religious ones, turn out to be devoid of meaning if they are analyzed according to the standards set forth by the aforesaid principle. What is, however, the status of the principle of verification itself? Is it an empirical sentence? Or, to put it differently, *how* is it possible to verify its truth or falsity using empirical observations? Obviously this cannot be done, and therefore we are bound to conclude that the principle cannot be submitted to empirical verification. But curiously—and this is the main point to be noted—the logical positivists meant to eliminate metaphysics using a principle whose nature, according to their own criteria, turns out to be metaphysical in character.

As we previously said, however, this fact is well known, and I promised instead to display some metaphysical tenets which, although endorsed by the neopositivists, are not usually taken into account. If we examine the famous distinction between *pseudo-problems* (which are, more or less, all those problems addressed to by traditional philosophy) and the *genuine* ones, it is easy to realize that, according to neopositivism, the difference between philosophy and science is the same difference holding between *language* on the one side, and *the world* described by language on the other. My thesis is that we can identify here a clear Kantian descent. Kant's work, while showing that metaphysics conceived of in scientific terms is impossible, linked science to the perceptual and conceptual characteristics of human experience. Acting that way, the philosopher of Königsberg hoped to avoid both the skeptical doubts put forward by David Hume and the metaphysical excesses often endorsed by the rationalists. It may be noted, however, that by limiting scientific discourse to a domain explicitly identified with *appearance*, Kant's writings prompted a growing interest in the *transcendent* domain which, if we take his words seriously, must exist somewhere beyond appearance itself.

So the neopositivists meant to solve this problem by rejecting the Kantian "synthetic a priori," and by reducing all knowledge to (i) purely empirical and (ii) purely linguistic factors, with nothing else left behind. The founder of the Vienna Circle, Moritz Schlick, claimed that between philosophy and science there is no conflict, but just a differentiation of their respective fields of inquiry: philosophy looks for *meaning*, and science is interested in *truth*. It follows that philosophers must only concern themselves with clarifying the meaning of scientific sentences, thus re-constructing the language of science in a clear and rigorous manner. Scientists, in turn, use language in order to ascertain the truth (or falsity) of those sentences concerning the world, and build theories which must be either verifiable or falsifiable.

What happens, then, if physicists want to discover the meanings of the assertions that are made within their discipline? According to neopostivism, if they do that, they become ipso facto philosophers. But on the other hand philosophers, in determining both the nature and the extension of meaningful discourse, set up the the boundaries of scientific inquiry, and this means, more or less, that philosophers themselves establish the *conceptual limits* of scientific inquiry. At this point nobody can deny that this is an extremely important job: the philosopher, in fact, becomes a kind of *superscientist*. He *bestows meaning*, and any operative scientist is practically compelled to ask for his—the philosopher's—opinion.

Let now ask ourselves, What does logico-linguistic analysis become *if* it is conceived of in these terms? The answer is that it becomes something which is enormously more important than the mere scrutiny of terms and sentences. It turns out to be a sort of *first philosophy*, a superdiscipline which is meant to establish the conditions that make all knowledge possible. If we, for some reason, do not want to call it "metaphysics," a different name may still be found for it. But it is quite clear that the substance of the argumentation does not change. And, in fact, in Wittgenstein's *Tractatus* we find the following claims:

> Philosophy settles controversies about the limits of natural science. It must set limits to what can be thought; and, in doing so, to what cannot be thought. It must set limits to what cannot be thought by working outwards through what can be thought. It will signify what cannot be said, by presenting clearly what can be said.[10]

If this is the situation, it becomes evidently misleading to speak of a "far" Kantian descent. Kant's presence in the neopositivist theoretical building is, instead, well-perceivable, and any difference must be attributed, in the final analysis, to the changes occurring in the historical context. So we find a first analogy between the Kantian reaction to rationalism and the neopositivist and analytic reaction to idealism. We also find a second analogy between Kant's desire to save scientific knowledge from Hume's skeptical doubts and the analytic attempt to assure safe logical foundations to that same knowledge. More generally, we may recall that for Kant the perception of reality was possible only if it were somehow *mediated* by conceptualization: our knowledge of the world always needs the application of categories, which in turn give shape to human experience. By adopting such an approach, it is no longer possible to speak of an *absolute* knowledge of reality, but only of a type of knowledge which is necessarily *relative to* our conceptual apparatus.

Exactly the same preoccupation is common to neopositivist and analytic philosophers, with only one important difference.

For Kant the conceptual apparatus filtering experience is located in human intellect, while the analytic philosophers of our century locate it in *language*. In both cases we have preconditions of knowledge: categories for Kant and language for the analytic tradition. So we are entitled to claim that contemporary analytic philosophy replaces the Kantian question *What are the conditions that make human knowledge possible?* with the following—and rather similar—query: *What are the conditions that make meaningful discourse possible?* Since the philosopher's task is just to answer this second question, the philosophical activity of bestowing meaning becomes—as Moritz Schlick used to claim—the beginning and the end of all knowledge. And it may even be noted that the problems foreseen by Kant about the possibility of mapping our conceptualizations onto reality-in-itself find a precise correspondence in the problems—underlined by neopositivist and analytic thinkers—concerning the possibility of mapping the characteristics of linguistic systems onto the reality they purport to name and describe.

But, at this point, the objections to the Kantian outlook are, *mutatis mutandis*, the same objections that can be addressed to linguistic philosophy. We can in fact claim that Kant, by posing limits to knowledge, assumed de facto the existence of something that lies beyond those limits. Similarly—as Wittgenstein, for instance, pointed out—to set up limits to meaningful discourse implies, ipso facto, assuming the existence of something that transcends those limits. We can thus conclude that, while according to Kant our knowledge of the world is relative to human conceptualization and categorization, for the analytic tradition this same knowledge, on a par with the meaningfulness of any discourse regarding the world, is relative to *language*. By choosing this path, conceptualization is transferred from human nature to language, and such a move is very important because it guarantees—theoretically, at least—the logical independence of the new linguistic philosophy not only from the old metaphysics, but also from natural science. This strategy favors, in particular, the detachment of philosophical inquiry from all kinds of psychological introspection, as Ludwig Wittgenstein underlines once again in his *Tractatus Logico-Philosophicus*:

> Psychology is no more closely related to philosophy than any other natural science. . . . Does not my study of sign-language

> correspond to the study of thought-processes, which philosophers used to consider so essential to the philosophy of logic? Only in most cases they got entangled in unessential psychological investigations.[11]

The antimetaphysical stance of linguistic philosophy is thus based on the assumption that our knowledge of the world is always relative to language and to our conceptual schemes, and the "Kantian flavor" of this position is neatly perceivable in a famous metaphor introduced by Otto Neurath, one of the founding fathers of the Vienna Circle. According to Neurath, in fact, we are all embarked since our birth on a sort of *conceptual ship*; if one wants to modify this ship, one cannot set ashore, but is rather compelled to rebuild it bit by bit in the open sea. This means that human beings cannot modify their conceptual schemes in order to make them more suitable to an—alleged—external reality, since this same reality (the world) turns out to be perceivable *only* through conceptual schemes. It is here that we find the roots of what I call "linguistic idealism," a theme about which we shall have more to say later on. It is not possible, therefore, to justify language by appealing to what reality is; adopting such a strategy, we are likely to return immediately to some old brand of metaphysics that, as we already pointed out, is for analytic philosophy a bunch of sentences devoid of meaning. And Neurath's metaphor is, as we shall see later,[12] very important for understanding the theses (or, at least, their early formulation) of a philosopher like Quine, who says in this regard:

> It is meaningless, I suggest, to inquire into the absolute correctness of a conceptual scheme as a mirror of reality. Our standard for appraising basic changes of conceptual scheme must be, not a realistic standard of correspondence to reality, but a pragmatic standard (on this theme see Duhem). Concepts are language, and the purpose of concepts and of language is efficacy in communication and in prediction. Such is the ultimate duty of language, science, and philosophy, and it is in relation to that duty that a conceptual scheme has finally to be appraised.[13]

It is correct to claim, at this point, that the purported analytic elimination of metaphysics is more theoretical than real. The absoluteness of traditional metaphysical questions like *What is the structure of reality?* or *Of what does it really consist?* has a punctual correspondence in the absoluteness of such linguistic questions as *What is the structure of our language?* or *What entities can we talk about?* The analytic tradition turned, in sum, the metaphysical absoluteness into a linguistic one, and this explains why Rudolf Carnap—just to take a famous example—meant to translate the traditional metaphysical and ontological assertions into sentences concerning the syntactic and semantic structure of language. And so, any sentence about *what there is* corresponds to a sentence about what *we say* there is. In Carnapian terms, the sentences of the metaphysical/ontological object-language must be translated into sentences belonging to the metalanguage of a philosophy conceived of as mere linguistic analysis.[14]

It is obvious, however, that this stance is nothing but the linguistic translation of metaphysical positions quite common in the history of Western philosophical thought. On the one hand, the possibility of pronouncing meaningful statements on an alleged extralinguistic reality is denied, but, on the other, we are supposed to pronounce definitive truths on the way in which we describe that same reality. According to this view, men would be unable to take directly into account the categories of existence, but would at the same time be able to examine in absolute terms their linguistic representations of reality. It is not difficult to understand that by acting that way the analytic philosophers, far from "eliminating" metaphysics, give rise to their own brand of metaphysics, with only one very important caveat: the world conceived of as reality, the Aristotelian being-*qua*-being, is replaced by Language (which must be written now with a capital *L*), thus going back to some form of rationalism. The alleged elimination of metaphysics through logical analysis of language is, therefore, a mere illusion, and it is much better to turn our attention instead to the symmetry existing between the analytic outlook and the positions endorsed by many traditional metaphysicians. No matter what the analytic thinkers claim, we have once again a distinction—in most cases *unconscious*—between science and metaphysics. As we observed previously, the analyst of language gives rise to a kind of first philosophy which is meant to bestow meaning on human knowledge conceived of as a whole. And thus a growing gap appears between *pratical-operative* scientific activity on the one hand, and linguistic analysis on the other. Abstract and purely speculative problems are privileged by many analytic philosophers, including traditional metaphysical and ontological problems which are now taken into account using the tools of formal logic and of the logical analysis of language. It is not at all surprising, thus, that the analytic thinkers have in most cases become the *linguistic counterparts* of those metaphysicians they originally meant to fight and destroy.

Within the analytic tradition, old distinctions are re-proposed which have always been objects of inquiry in traditional metaphysics: for instance the distinction between platonists and nominalists. The clarification of the traditional philosophical problems obtained thanks to the linguistic turn, in other words, does not succeed in hiding the much more important fact that many old quarrels show up again within the analytic field. Let us only mention the dispute about the opportunity of using a nominalist or a platonist language in the philosophy of logic and mathematics. This is not a merely linguistic controversy but, rather, the re-proposal of problems thriving already in ancient and medieval philosophy. The step toward recognizing that metaphysics, after all, cannot be eliminated is rather short.

Carnap's answer was that the analytic discussions about platonism and nominalism do not regard the problem of the existence of universals, but rather the choice of which language is more useful for dealing with themes related to the foundations of mathematics. It is evident, however, that this answer is far from being satisfactory. Platonists and nominalists fight each other not for linguistic reasons, but because they do not agree on what kinds of entities have to be admitted into *ontology*. The real nature of the dispute is ontological and metaphysical, and not linguistic. So the antimetaphysical stance turns out to be more an "ideological prejudice" (just to borrow a term belonging to political philosophy) than a thesis which can be reasonably argued for.

It may then be noted that the absolutism of the analytic conception of language, which is based on the thesis that our talk about the world is meaningful only in so far as it is referred to some system of linguistic representation, practically assumes that language itself is *not* a part of the world. In other terms, we must ask a question that neopostivist and analytic philosophers tend—strangely—to ignore: *How* was language born and *where* does it come *from*? This is the reason why Quine (or, at least, the Quine of the 1950s) rejects the analytic/synthetic distinction, insisting instead on language conceived of as a *tool* created by mankind for practical purposes.[15] The early Quine, thus, overcomes the strictures of a purely analytic conception of language by resorting to the pragmatist—and typically American—tradition represented by thinkers like James, Peirce, and Dewey.

Let us conclude this section by noting that, despite what many authors nowadays keep saying, (i) metaphysics has not been eliminated and (ii) logical analysis of language may instead be extremely useful for examining the traditional metaphysical problems in a more precise manner. This explains why, within the analytic school itself, a great deal of attention is currently paid to such old problems as the platonism/nominalism distinction, the status of modal propositions, and the meaning of negative existential sentences.[16] In any event, our criticisms are not meant to overlook the fact that neopositivism and analytic philosophy have great merits too, because they compelled philosophers to take eventually into account the scientific dimension of human inquiry. Their work, furthermore, is an important step toward the construction of a philosophical language which is both rigorous and intersubjective.

5. THE POST-EMPIRICIST TURN

It is well-known that, in the last few decades, the neopositivist supremacy within philosophy of science has been replaced by that of the so-called post-empiricist thought. My intention is not to outline an historical account of this important change of outlook, because this has been done already by other authors.[17] What I want to do, instead, is to give some short and sketchy remarks on the post-empiricist phenomenon, always taking into account the themes of the relationships between language on the one side and reality on the other, and the parallel problem of the alleged elimination of metaphysics. Let us begin with Karl R. Popper, whose falsificationism, although from some aspects still close to neopositivism, nevertheless gave rise to a true epistemological revolution.

Unlike the neopositivists, Popper believes that a clear separation (i) between analytic and synthetic sentences and (ii) between theory and observation is an impossible task; according to his view, a general sentence is not produced by many particular observations, and scientific theories are nothing but *conjectures* freely created by the human mind in order to explain empirical phenomena. Theory and observation are thus intimately linked to each other, and no pure—and detached from some theoretical context—observation is ever possible. A very important point may be noted here. A position very similar to Popper's was endorsed by the American pragmatists, in the last century with Charles S. Peirce and—especially—William James, and in the present one with John Dewey. To my knowledge this fact is not very well known nowadays, maybe because the analytic thinkers usually do not pay much attention to the history of philosophy. Only recently, for instance, did Hilary Putnam rightly underline in a volume published in Italy these beautiful remarks by James concerning the relationships between theory and observation:

> The knowing subject is no mirror reflecting passively an order that already exists. The knowing subject is an actor, who both codetermines truth and registers the truth he manages to create.[18]

Putnam, however, does not mention the striking similarities between Popper's and James's theses, although we cannot infer from this fact that the founder of falsificationism was somehow influenced by the American pragmatist tradition. Anyhow, it is clear that if we recognize that the theoretical dimension precedes observation, and if we claim furthermore that scientific theories have a creative character, then we may explain the "jumps" that often take place in the history of science (i.e., the genial intuitions that allow scientists to interpret usual phenomena in a new way) better than by having recourse to the classical neopositivist model.[19] Einstein himself, in fact, used to say that there is no logical and safe path able to take us automatically to the discovery of the universal laws of physics, since only a mixture of intuition and experience may bring scientists in the right direction.

In criticizing neopositivism, Popper says something new and important about the relation between language and the world. In the 1959 foreword to his most famed epistemological work, in fact, he claims:

> Language analysts believe that there are no genuine philosophical problems, or that the problems of philosophy, if any, are problems of linguistic usage, or of the meaning of words. I, however, believe that there is at least one philosophical problem in which all thinking men are interested. It is the problem of cosmology, the problem of understanding the world—including ourselves, and our knowledge, as part of the world. All science is cosmology, I believe, and for me the interest of philosophy, no

> less than of science, lies solely in the contributions which it has made to it. . . . It seems to me paradoxical that philosophers who take pride in specializing in the study of ordinary language nevertheless believe that they know enough about cosmology to be sure that it is in essence so different from philosophy that philosophy cannot make any contribution to it.[20]

True, according to our author this is "cosmology," but would the meaning of his words change by using, instead, the term "metaphysics"? I do not believe so, even though, as I said previously, that term may be abandoned if for some reason it is deemed to be bothersome. It is in any event obvious that a mere change of words does not modify the substance of things. Popper thinks that "prohibiting" any talk about nonscientific problems is simply ridiculous; men's talk is both scientific and nonscientific, and, far from eliminating one of the two, it is instead important not to confuse them.

It follows that, according to the Austrian-born philosopher, metaphysics *is* a part of meaningful discourse, even though metaphysical theories are different (i.e., they are not falsifiable) from scientific ones. But there is a sort of contiguity between science and metaphysics, because we often find in the history of thought metaphysical theories which, later on, became scientific. The real problem at stake is, thus, to find valid criteria of demarcation. Since metaphysical ideas are often able to influence scientific activity, metaphysics turns out to have a precious *euristic* value, and any attempt at eliminating it is bound to failure.[21] Using Ian Hacking's words, we may say that:

> He [Popper] does define science as the class of testable propositions, but far from decrying metaphysics, he thinks that untestable metaphysical speculation is a first stage in the formation of more testable bold conjectures.[22]

Popper, however, is still close to neopositivism when he claims that the metaphysics' role is after all auxiliary, because it is only within science that men can get a true knowledge of reality.

Turning our attention now to the "methodological anarchism" endorsed by Paul K. Feyerabend—and leaving aside, for reasons of space, any reference to such important authors as Thomas Kuhn, Imre Lakatos, and many others—we may note that, today, the role played by science within the vaster domain of human knowledge has been strongly questioned. Feyerabend, for instance, claims that man's landing on the Moon has no objective value. This event is in his view important only to the citizens of the Western and industrialized nations, who are accustomed to judging reality through the lenses of scientific progress. In his opinion the shaman of a primitive tribe has a kind of knowledge which is by no means inferior to that of Western scientists, while myth and science are on the same level: they are just different world-perspectives, which are both valid and endorsable. Taking this path, the notion of scientific progress becomes meaningless: if someone believes that magic is better and more useful than science, let him think that way. And if we underline the *practical* results obtained by science, Feyerabend's answer is that there is no reason to judge the goodness of a particular world-perspective on practical bases. At this point, no doubt, any kind of dialogue becomes rather difficult. Feyerabend's theses mirror ideas and cultural tendencies which are nowadays popular with public opinion. Newspapers and TV programs pay great attention to all kinds of magic and esoteric doctrines, and Feyerabend's popularity is thus understandable if we take it as a typical sign of our times.

What, then, is the role of philosophy of science according to Feyerabend? And what is the role of philosophy at large? Such a role does not even exist, we might answer. Note that, in reacting against the typical scientistic outlook endorsed by neopositivism, Feyerabend is—at least partially—right. We may well agree that science, after all, does not give us the only type of knowledge worth pursuing. But is it necessary, when claiming that there are many kinds of knowledge available to human beings, to deny the fact that science—and *not* magic—gave us a fair knowledge of the structure of the solar system? Feyerabend's answer, "Who cares?" makes any rational discussion pointless, and even questions the classical definition of man as an animal who naturally aims at knowing the reality which surrounds him. For this reason, we should not be surprised when reading these statements drawn from a recent work of his:

> "To be a philosopher" either means that we approach the world as members of a club, or is a void expression, which can be applied to any individual, even to a dog. I gladly admit to be a philosopher in the second sense of the term but, certainly, not in the first one.[23]

Nobody can thus deny that Feyerabend endorses irrationalism (a charge that our author, after all, does not reject). But in my opinion a philosopher of science is not entitled to assume such a stance, because science is, by definition, a rational activity. So we may conclude that Feyerabend is no longer a philosopher of science, but something else: maybe a poet, or a propagandist of magic and esoterism. This is, according to my view, one of the main reasons why contemporary scientists no longer take philosophy of science seriously. Some philosophers answer that Feyerabend's positions, far from being irrationalist, are instead justified because they take into account a *different* kind of rationality. To which I reply: "All right, but *which* one?" His rationality might be valid in Terry Brooks's fantasy stories, which are certainly amusing and well written, but which do not reflect reality at all. In everyday life, unfortunately, we have no magic swords, no elves, and no knights endowed with supernatural powers who fight the forces of Evil.

We do not need to accept Feyerabend's theses in order to reject the scientistic outlook on reality endorsed by the neopositivists, or the linguistic absolutism put forward by the analytic conception of language. A rediscovery of American pragmatism is more than sufficient for that purpose. We may find that pragmatism and neopositivism share many common positions: both are interested in science and its methodology; both have faith in human reason and its capacities; both believe that philosophers must try to *demonstrate* in a rigorous manner the correctness of their statements, giving rise to an intersubjective type of discourse. The neopositivists, however, endorse scientism and the pragmatists do not. According to pragmatism, scientific knowledge is just one of many, available kinds of knowledge (although being very important and central), while for logical positivism no meaningful talk is possible outside the scientific field: every type of knowledge must be reduced to the scientific one. So we have monism and reductionism on the one side, and pluralism and antireductionism on the other. The pragmatists, unlike the neopositivists, stress the primacy of *practice*, a fact on which any responsible scientist may easily agree. The analytic philosophers—and especially Carnap—believe in the existence of *one* method able to solve all epistemological problems, and this method is essentially based on the tools furnished by formal logic. Dewey, however, realized well in advance that this was a *philosophical utopia*,[24] and now we may well understand the pragmatist rejection of the excesses of logical formalism. That rejection seemed outdated for a long period of time, while today it looks more and more justified. And what about the relationships between science and ethics? While Dewey deemed them essential, the neopositivist and analytic philosophers usually paid little attention to this topic.

So there are more rational ways for rejecting scientism than the endorsement of astrology and magic. Yet, many people claim that Feyerabend's positions are a necessary and inevitable reaction to neopositivism's scientistic dogmas. In philosophy as anywhere else, they say, one excess always prompts an opposite one. When someone reduces rationality *as such* to scientific rationality, it is not surprising to find authors who claim that the landing on the Moon has no objective value. This thesis, however, is too strong: we should not talk of "inevitable reaction" but, rather, of the creation of a peculiar cultural climate. No doubt logical positivism and analytic philosophy have some responsibility if we live today in a cultural environment that largely accepts an irrationalist outlook on reality. If we claim that science is *everything*, we are likely, sooner or later, to be confronted by someone objecting that science is, instead, *nothing*. When common men find out that science cannot face all problems, they will most likely turn their attention to charlatans who promise to solve all those problems science is unable to cope with.

Let us conclude our analysis going back to Feyerabend's theses once again. He attacks—and this is indeed an original stance—scientific rationality *from inside*, by using epistemological tools. Philosophy of science, thus, finds enemies within the walls of its own stronghold: Feyerabend and his followers have turned philosophy of science into something mysterious and not easily classifiable in philosophical or scientific terms. The *anything goes* undermines the meaning itself of the discipline, to the point that scientists find it difficult to understand what Feyerabend's philosophy of science is up to. If science is equated to any other dimension of spirit—art, religion, or even witchcraft—the specific and cognitive character of scientific rationality is eliminated. It follows that philosophy of science loses any meaningful role within the field of human knowledge, while even philosophy as such becomes more similar to a joke than to a serious endeavor. In Feyerabend's opinion, both philosophy and the philosophy of science are nothing but products of our particular—i.e., Western—kind of rationality, which has no supremacy on different types of rationality, including those endorsed by primitive tribes. He thus claims that history must be defended by freeing it from the (alleged) chains that Western epistemology has placed everywhere, reevaluating all the different cultural traditions. It seems to me that the physicist Steven Weinberg gives the right answer to such arguments:

> I suspect that Gerald Holton is close to the truth in seeing the radical attack on science as one symptom of a broader hostility to Western civilization that has bedeviled Western intellectuals from Oswald Spengler on. Modern science is an obvious target for this hostility; great art and literature have sprung from many of the world's civilizations, but ever since Galileo scientific research has been overwhelmingly dominated by the West. This hostility seems to me tragically misdirected. Even the most frightening Western applications of science such as nuclear weapons represent just one more example of mankind's timeless efforts to destroy itself with whatever weapons it can devise. Balancing this against the benign applications of science and its role in liberating the human spirit, I think that modern science, along with democracy and contrapuntal music, is something that the West has given the world in which we should take special pride. In the end this issue will disappear. Modern scientific methods and knowledge have rapidly diffused to non-Western countries like Japan and India and indeed are spreading throughout the world. We can look forward to the day when science can no longer be identified with the West but is seen as the shared possession of humankind.[25]

Granting that Feyerabend is certainly right in saying that different cultural traditions have the right to be respected and preserved, it must nevertheless be noted that nature does not react in the same way to any kind of inquiry. This is the real point at stake: some actions are successful and others are not, and this means that reality itself poses *structural* limits and bounds to human research, limits and bounds that men cannot overcome. Feyerabend's relativism misses the point precisely for this reason.

These are the roots of the growing conflict between philosophers of science and practicing scientists. Despite any difference of opinion, the dialogue between a neopositivist or a Popperian philosopher of science on the one side and a professional scientist on the other is always possible. But when it comes to Feyerabend and his followers, this dialogue turns out to be null. The current situation is thus dangerous, because it tends to deepen a gap between philosophy and science that was—at least partially—filled in the first decades of our century.

CHAPTER 2

Between Language and Reality

1. QUINE AND ONTOLOGY

If there is at all an analytically oriented philosopher who often writes about ontology, it is Quine. Throughout his writings, in fact, we constantly find the following question: *What kinds of objects do exist?* On the other hand, the so-called ontic decision—that is, the decision concerning what objects should be admitted into ontology—is the cornerstone of his philosophical quest. A central position has been given to these problems in Quine's works since the late 1930s, and important formulations may be found in such writings as "On What There Is," *Word and Object*, and "Ontological Relativity." Given this fact, one would be inclined to think that in those works a precise and well-constructed ontology can be found. The real situation, however, does not meet these expectations.

Quine firmly believes that ontological problems can be addressed (and eventually solved) *only* by means of symbolic logic. But one has to be very careful in pronouncing such a statement. I do not mean to deny that formal logic is a very important tool when ontological questions are coped with, since this fact may be easily recognized by anyone trained in analytic philosophy (an existentialist would not agree, of course, but this fact is not important for our purposes). In Quine's view the above-mentioned statement is much stronger, since he claims that ontological problems *are*, at bottom, logical matters: standard quantification theory, and in particular existentially quantified sentences, are in his opinion the ontological idiom par excellence. It follows that, should we not possess first order logic, we would not even be able to express ontological questions properly.

This is indeed a tough-minded position, and it takes us back to the relationship between *language* and the *world* (which is, as everybody can easily understand, the most serious ontological problem within the analytic tradition). Why does Quine assume such a stance? In order to answer this question, one must be audacious enough to challenge Frege's paradigm, which identifies *existence* and *logical quantification*, a challenge that very few authors today feel inclined to make.[1] In fact, Quine is one of the champions of this identification of existence with standard logical quantification, a very important move indeed, since it was—and still is—endorsed by the absolute majority of the authors belonging to the analytic tradition. So, if we want to answer the question posed above, I think that there are two possible replies, which are in turn interconnected:

A. Only formal logic can adequately describe reality, and
B. It is impossible to make a distinction between logic, language, and reality.

But the real question at stake here is no doubt the following one:

C. Is there a world (reality) independent of the language we use for *talking about* reality?

Let us note that Quine underlines, here and there in his writings, that this—i.e., (C)—is precisely the problem we have to solve in order to get a satisfactory comprehension of what ontology is and what it means to formulate ontological questions. And in fact, when posing existential problems in his works, Quine always adopts this formulation:

D. May we use a term x without presupposing, by that same fact, the *existence* of the entity (object) named by x?

In other words, what worries him is the possibility that a speaker uses nondenoting or general terms, just believing that there is something real named by them. But—we may ask—are those really *ontological* problems? It seems to me that they are, instead, *logical* and *linguistic* ones. For example, I may find someone believing in the existence of winged horses. If he tells me that Pegasus exists, in order to show him that this statement is incorrect I will not use quantification theory but, rather, I invite him to pursue an empirical recognition. It may turn out, at some point, that he believes in the existence of mental entities, thus rendering a direct empirical recognition quite hard. Even in that case, though, we may agree on the fact that mythological texts report many stories about Pegasus and winged horses. Provided I am ready to accept an extended ontology including all kinds of mythological objects, I may also concede that the sentence *Pegasus is the winged horse of Bellerophon* is true, while the sentence *Pegasus is the winged horse of Hercules* is false. I may subsequently formalize those two sentences using first order logic, but this move will neither solve nor add anything to the question of whether there *really* are winged horses or not. The solution, in fact, depends on two facts:

1. My interlocutor's ability to find winged horses in the empirical world; and
2. His (and my) background ontological assumptions; if he is an empiricist, he will make one claim, and if he is a mentalist he will make yet another claim.

The fact of the matter is that mathematical logic allows us to formalize the two sentences in a clear manner but, on the other hand, offers no solution to the philosophical problem we are interested in, since the existential quantifier (or particular, if we adopt a substitutional interpretation of the quantifiers) is a purely *formal* tool and not an ontological one. The same applies to scientific reasoning: certainly first order logic does not tell us whether non-observable entities really exist or not (that is, the fact that we talk about them is no guarantee of their existence).

Going back to Quine's theses, it is difficult to find in his writings a clear distinction between what is logical (linguistic) and what is empirical. He certainly did write a lot about what there is, but never gives us a chance to find out whether, in his opinion, there really is a world independent of the language we use for *talking about* the world (and we will see later that sometimes he seems to endorse, in this regard, rather incompatible positions). Language, thus, assumes an ontological dimension that is not acceptable if we take language to be not an a priori element but rather a socially, historically, and culturally determined *instrument* which merely refers to reality (and a very imperfect instrument, too). As science shows, man came first and language later.

I think that if we follow Quine's style of reasoning we are not in the domain of ontology (whose boundaries are given by the search for what there is), but rather in the domain of formal logic, and the two domains, notwithstanding the opinion of many important analytic philosophers, are in my opinion by no means identical. If we pose the previous questions in a Quinean style, all we can do is to find out that nondenoting singular terms like *Pegasus* or general terms like *centaur* do not work—in language—as usual names do. No doubt this is an important result, but its consequences regard formal logic and the philosophical analysis of ordinary language, and not ontology.

It is thus difficult to understand how one may determine the existence (or the nonexistence) of certain objects when adopting Quine's point of view. We can find a possible solution to this puzzling problem if, and only if, the existence of an external reality independent of any knowing subject is questioned. It would therefore be necessary to adopt a *different* notion of reality, inextricably connected with language and such that its meaningfulness coincides with our capacity to express it at the linguistic level. And in fact he gives, here and there, hints in this direction: sometimes he claims that "ontology is *internal to* language." But at this point we must underline a very important fact: Quine does not even seem to be conscious of the just-mentioned implication which logically follows from his treatment of the ontological problems. Indeed we find, here and there in his essays, scattered remarks on the relationship between language and the world. None of these remarks, however, is clear and precise enough to offer a sound foundation for the existential questions. Quine looks satisfied with his logical formalizations and does not try to go deeper in the purely philosophical background of his theses.

This remark of mine may seem at first sight a little strong, but it is supported by a careful analysis of Quine's texts. Let us note, by the way, that our philosopher always claims that he is a convinced empiricist, and this is the reason why he is so often prisoner of two alternatives, which seem incompatible:

1. On the one side, he supports a strong version of empiricism, according to which reality is simply (and *only*) what we can have experience of from the sensory viewpoint; but,

2. on the other, he supports a sort of logical and linguistic realism, according to which reality is simply (and *only*) what we are able to express within our language adequately modified by recourse to logical formalization (first order logic).

Quine never succeeds in clarifying this important issue: he does not even try to, probably because he does not have a clear perception of the difficulty of putting together the two alternatives just mentioned. His neopositivist legacy takes him to a form of radical empiricism, while his logical realism brings him in quite a different direction. I know, of course, that much the same can be said about many representatives of logical positivism itself, but the analysis of the "hidden metaphysics" of logical empiricism—a fascinating theme indeed—is not our main concern in this context, although it can be reasonably argued that it is important for analyzing Quine's theses, too. Let me just say that, despite their official anti-metaphysical attitude, logical positivists created one of the most beautiful metaphysics of our century.[2]

In any event, only by taking this inner tension into account is it possible to make sense of Quine's claim that Pegasus does not exist because the term *Pegasus* does not work as a real name when it is encapsulated in a sentence. It follows—in his opinion—that we must resist the temptation to adopt an "overcrowded" ontology . But I must stress that the attempt to base ontology on this sort of *linguistic/aesthetic* preference looks very strange: if reality were truly overcrowded from the ontological viewpoint, would it follow—according to Quine—that we should give priority to his (or even our) linguistic/aesthetic preferences rather than to the ontological structure of reality itself? Maybe Ockham's razor, after all, is not such a good ontological criterion: why should we decide to simplify, following our personal taste and opinions, a reality which is itself complex? The fact of the matter is, in my view, that we must make our ontological decisions on *empirical* bases, and not on logical and linguistic ones.

What objects, then, are we allowed to admit into our ontology? It is clear, in fact, that at least some objects do exist. Since according to Quine we cannot admit possible objects, meanings, intensional and abstract entities, what remains to be done is to determine the scope of our *ontological commitment*. Quine's answer is the following: the existence of any object is *relative to* the language we speak, and is therefore *based on* such a language. If this is the situation, it would seem to follow that there are no objects independent of the language we use to talk about them and that, furthermore, there are no objects independent of the linguistic terms which denote them (although, as we shall see later, in many occasions Quine seems to hold a different position). Physical objects become postulated entities that help us organize the flux of experience. However popular these theses are in contemporary philosophy (and not only within the analytic tradition: let us think, for example, of hermeneutics and philosophers like the late Heidegger, Gadamer, Apel, Habermas to a certain extent, etc.), I must confess that I cannot help but consider them ill-founded.

Furthermore, I firmly believe that we can formulate the following question: is this—Quine's—an appropriate use of the term *ontology*? I do not think so, because the notion of an ontology based exclusively on language is in my opinion misleading for this simple reason: language has an intrinsically *referential* character,[3] so that it can never be thought to be, just by itself, the foundational basis of ontology (in other terms, we do not learn language just to allow our conceptual scheme to talk to itself). The solution to all these problems might be found, perhaps, substituting the word *ontology* with the expression *logical universe of discourse*, but Quine never makes such a move, and so we are back once again to the alternative previously mentioned:

E. Either language and reality are the same thing, or

F. Language and reality, even though they do not coincide completely, are so inextricably connected that it is impossible to distinguish one from the other.

No wonder, then, that when Quine writes or talks about ontology, he always refers to language. Since the existence of objects is determined by the language one speaks, any decision concerning ontological problems implies the clarification of which ontological commitments are embedded in a particular language, and this, in turn, means that speakers of different languages assume (or may assume) *different* ontological commitments, so that ontology is internal to language as we said before. Notice, by the way, that at this point we begin to identify the roots of Quine's influence on the current scientific anti-realism and, more generally, the roots of the negative influence that philosophy of language today exerts on philosophy of science. The alleged incommensurability of scientific theories, in fact, stems also (and it may be said even "mainly") from such ideas.

Ontology, thus, can be adopted according to our particular preferences. Reality poses no burdens and no limits, and any ontic decision is a linguistic one (that is, linguistic in character). It is not an alleged objective reality that tells us what ontology to adopt, but it is instead the language we speak that plays this fundamental role. True, Quine claims that the adoption of an ontology obeys pragmatic requirements but, still, it is *language* that takes us in this direction.

Now, we should at this point ask ourselves, What kind of ontological status—if any—would objects have in a world identical to ours except for one, very important, characteristic: the total absence of mankind? Notice that I am not talking of a "possible world" in the—sometimes—weird sense attributed to this term by contemporary modal logicians, but of a *real* world; it might be, for instance, our Earth before men appeared on its surface, or our planet following a terrible nuclear war in which the whole of mankind was destroyed (so, we do not even need the exotic dreams of some modal logicians to imagine a world without mankind). But astronomic and cosmological research—that is, operative science—also shows us many examples of planets which developed without the presence of forms of life equivalent to human beings.

Quine would most likely answer that such a question is irrelevant: since there is no human subject in that context, it does not even make sense to ask for the ontological status of that world and the objects it contains. And the same answer would most likely be given by many linguistically oriented philosophers of science. Note that even an existentialist would give a similar answer but, given the premises from which he starts (only human existence is important), his reply is surely more justified than Quine's. But this is precisely what I want to deny as firmly as possible because it denotes a wrong-headed way of approaching both the world and philosophy, besides being intrinsically antiscientific. Quine always talks of *scientific philosophy*, and he is certainly right in this respect, because a sound philosophy cannot ignore, nowadays, the scientific dimension of human knowledge. And this is even more correct because science is the paradigm of knowledge, even though, saying this, I do not mean to uncritically adopt a logical empiricist style of doing philosophy.

On the other hand, however, you will not find many scientists who link so strictly objective reality to our subjective capacities and, especially, to our linguistic competence. So, since we are dealing with the influence of philosophy of language (and formal logic) on today's philosophy *of science*, let us see what two professional scientists think on this subject.

The physicist Stephen Hawking, for example, in the final chapter of his best-seller *A Brief History of Time: From the Big Bang to Black Holes,*[4] makes the following remarks: How come—he wonders—contemporary philosophers are only concerned with linguistic analysis? Don't they realize that, acting that way, they cannot understand what scientists are really interested in (even though—we may add—many of them think they are doing philosophy of science)? These are Hawking's exact words:

> Up to now, most scientists have been too occupied with the developments of new theories that describe *what* the universe is to ask the question *why*. On the other hand, the people whose business is to ask *why*, the philosophers, have not been able to keep up with the advance of scientific theories. In the eighteenth century, philosophers considered the whole of human knowledge, including science, to be their field and discussed questions such as: Did the universe have a beginning? However, in the nineteenth and twentieth centuries, science became too technical and mathematical for the philosophers, or anyone else except a few specialists. Philosophers reduced the scope of their inquiries so much that Wittgenstein, the most famous philosopher of this century, said, "The sole remaining task for philosophy is the analysis of language." What a comedown from the great tradition of philosophy from Aristotle to Kant![5]

Obviously we do not have to agree with Hawking's vision of philosophical work, but, no doubt, he underlines a very serious shortcoming. And partly analogous comments may be found in Steven Weinberg's latest book:

> Physicists get so much help from subjective and often vague aesthetic judgements that it might be expected that we would be helped also by philosophy, out of which after all all our science evolved. Can philosophy give us any guidance toward a final theory? . . . The insights of philosophers have occasionally benefited physicists, but generally in a negative fashion—by protect-

> ing them from the preoccupations of other philosophers. . . . In our hunt for the final theory, physicists are more like hounds than hawks; we have become good at sniffing around on the ground for traces of the beauty we expect in the laws of nature. . . . Physicists do of course carry around with them a working philosophy. For most of us, it is a rough-and-ready realism, a belief in the objective reality of the ingredients of our scientific theories. . . . This is not to deny all value to philosophy, and I do not even mean to deny all value to the philosophy of science, which at its best seems to me a pleasing gloss on the history and discoveries of science. But we should not expect it to provide today's scientists with any useful guidance about how to go about their work or about what they are likely to find.[6]

We may note that, today, many contemporary philosophers of science could not even make sense of most of these words. Certainly they deem more or less useless the search for a final theory,[7] while the best scientists seem to be engaged in just that kind of research. The thesis of the alleged incommensurability of scientific theories, moreover, was raised in the philosophical circles and has had an enormous impact on philosophy of science, while most scientists do not even seem interested in it. The reason is very simple and perfectly understandable: were scientific theories really incommensurable, and were scientists really convinced that this thesis is correct, there would be no science left. We would have had that "death of science" that many antiscientific philosophers nowadays seem to invoke, following, for example, Martin Heidegger's theses (which makes his growing popularity a little frightening). But, of course, we all know that the actual situation is different, and scientists claim they are getting closer and closer to what they call the "final theory," that is, the discovery of the final laws of nature (so Steven Weinberg argues in the book I have just quoted, and Stephen Hawking says more or less the same things; we shall speak again later on of this big split between philosophy of science on the one side, and operative science on the other, an argument with which we already dealt briefly in the preceding chapter).

In any event, most likely an astronomist looking for far-distant galaxies would not even dare to claim that the ontological status of those galaxies (and of the worlds contained in them) depends on what our human language allows—or does not allow—us to say. Likewise, a paleontologist who does scientific research on the world before mankind's appearance on the Earth would not even dare to claim that the objects of that world do not have an ontological status independent of our conceptual schemes. He will look, instead, for those objective data that may explain how that world really was and looked like. If we reported Quine's statements to those two scientists, they would probably say, "Well, if *this* is scientific philosophy or philosophy of science nowadays, we find it neither interesting nor useful to our work." Notice that this is exactly what Hawking and Weinberg say in the previous quotations, and it would be very easy to find other examples of the same kind.

As a matter of fact, some contemporary philosophers of science have come to realize that philosophy of language (and formal logic) cannot have such a prominent importance for their discipline. Mario Bunge, for instance, when dealing in a book of his concerning causality with the possibility of founding a scientific metaphysics, explicitly denies that philosophers of language have something to say in this respect.[8] The same position is endorsed by Wesley Salmon,[9] while Ian Hacking, when talking about the concept of "observation," argues this way: "Commonplace facts about observation have been distorted by two philosophical fashions. One is the vogue for what Quine calls semantic ascent (don't talk about things, talk about the way we talk about things). The other is the domination of experiment by theory. The former says not to think about observation, but about observation statements—the words used to report observations. The latter says that every observation statement is loaded with theory—there is no observing prior to theorizing."[10] And he goes on, claiming:

> Seeing is not saying. The drive to displace observations by linguistic entities (observation sentences), persists throughout recent philosophy. Thus W. V. O. Quine proposes, almost as if it were a novelty, that we should "drop the talk of observation and talk instead of observation sentences, the sentences that are said to report observations" (*The Roots of Reference*, pp. 36–39). Caroline Herschel not only serves to rebut the claim that observation is just a matter of saying something, but also leads us to call in question the grounds for Quine's assertion. . . . There is, Quine says, a perfectly distinguishable class of observation sentences, because "observations are what witnesses will agree about, on the spot." He assures us that "a sentence is observational insofar as its truth value, on any occasion, would be agreed to by just about any member of the speech community

> witnessing the occasion. And we can recognize membership in the speech community by mere fluency of dialogue." It is hard to imagine a more wrong-headed approach to observation in natural science. . . . Quine's agreement "on the spot" has little to do with observation in science. If we want a comprehensive account of scientific life, we should, in exact opposition to Quine, drop the talk of observation sentences and speak instead of observation. We should talk carefully of reports, skills, and experimental results. We should consider what, for example, it is to have an experiment working well enough that the skilful experimenter knows that the data it provides may have some significance.[11]

Bunge, Salmon, and Hacking, however, today are less fashionable than linguistically oriented authors. Most philosophers of science continue to hold that scientific theories are *only* sets of linguistic statements (while I think they are also—but not only—sets of linguistic assertions), and that coherence among such statements is the only important requisite to look for. Note again, as we already said before, that such an approach seems to miss the essentially *referential* dimension of human language (assuming, of course, that we are ready to endorse a vision that takes language to be just one of many human products, and not a sort of inexplicable a priori entity). And let us note, furthermore, that it is not correct to take into account—as Quine does—only language (words) on the one side and human behavior on the other. We need something else to explain how our language really works. Assuming observable behavior (together with language) as the absolute starting point, we will soon run into the question of the other minds, and this is exactly the problem Quine meets in developing his famed doctrine of the indeterminacy of translation (this point will be taken up again in the next section).

Quine in fact believes that, when we translate theoretical propositions, the totality of the physical facts does not determine our translation. This may be correct, but we should also ask him: Why do you think that only *behavioral* facts—which after all are only one particular subset of the set of physical facts—are important for our translation of theoretical sentences? In any event, if we pursue this exclusively linguistic style of reasoning, we are not likely to understand concrete scientific practice. The final outcome will be a progressive and dangerous split between practicing scientists on the one hand and philosophers of science (or, at least, those of them who adopt an exclusively linguistic approach) on the other (but this already is the case). And what kind of usefulness may a philosophy of science ever have which looks interesting to philosophers of language and logicians, but *not* to scientists themselves? It seems to me that all philosophers of science should seriously reflect on this fact. If Weinberg entitles the seventh chapter of his aforementioned book "Against Philosophy" (and the real target is philosophy of science), philosophers of science should at least try to understand why this happens.

Furthermore, I do not agree with Hacking when he writes about "philosophical fashions" that have determined such a situation. It is not only—and not mainly—a matter of fashions, because we should instead talk of the prevalence of the linguistic turn within philosophy of science. As anyone may easily understand, it is easy to abandon fashions, but it is instead difficult to fight against a dominant paradigm. The battle will probably become easier only when new paradigms take over, and there are indeed signals that this could happen in a relatively short time (even though, I must say, I do not feel comfortable seeing analytic philosophy replaced by Heideggerianism and hermeneutics). It is clear, anyhow, that after decades of semantic ascent, philosophy of science needs now a semantic *descent* in order not to lose any contact with concrete scientific activity.

2. THE MYTH OF ONTOLOGY INTERNAL TO LANGUAGE

Let us now go back to our previous remarks on the language/world relationship according to Quine, because it is a fact that the situation depicted in his works is not as clear as I have thus far decribed it. Despite his deep interest in ontological problems, in fact, Quine's works sometimes present statements that do not seem easily reconcilable. In an important essay entitled "Existence," our author makes the following claims:

> Which ontology to ascribe to a man depends on what he does or intends with his variables and quantifiers. This second appeal to language is no more to be wondered at than the first; *for what is in question in both cases is not just what there really is, but what someone says or implies that there is. Nowhere in all this should*

> *there be any suggestion that what there is depends on language.*[12]

It should be noted that, according to Quine, formal logic determines ontology: there is no other way to make sense of his statement that "which ontology to ascribe to a man depends on what he does or intends with his variables and quantifiers." And this means, moreover, that ontology is relative to language. Subsequently, however, I underlined the last two sentences of the quotation because I think that, in this case, the distinction between reality and language is pretty clear or, at least, it is clear that Quine is *not*—in this context, at least—identifying reality and language. This means however that, if we adopt his strategy (which, as we said earlier, is based on standard quantification theory), we remain at the linguistic level without touching the properly ontological one, in the sense that we just cannot step outside the barrier posed by our language; and, in fact, Quine's ontological commitment is not about what there is, but about what a true theory *claims* that there is. Note, by the way, that I often add the adjective *human* to *language* just to stress that language is a historically determined human product. In any event, Quine in the essay mentioned above recognizes, although confusedly, that *there is* an ontological level which cannot be identified with the linguistic (logical) one. This is confirmed by what he claims soon afterwards:

> What are we to say of the ontic commitment of a New Guinea aborigine—what are the purported values of his variables? I hold that there is in general no unique translation, not even unique in respect to ontic commitment, let alone logical style. I hold that our distinctively referential apparatus. . . . belongs in an essential respect to the theoretical part of our language: namely, it is underdetermined by all possible sensory stimulation. A result, or really another way of stating the point, is that our referential apparatus is subject to indeterminacy of translation. That is, translations not equivalent to each other could be reconciled with all behavior. . . . It may in this sense be said that ontological questions are parochial to our culture. This is not to say that a thing may exist for one culture and be non-existent for another. *Existence is absolute, and those who talk of existence can say so*. What is parochial is the talking of it.[13]

So—I think we may conclude—after all there is an external reality which does not depend on language. Let us see, then, if it is possible to detect a definition of it in Quine's works. We can find there two different concepts of reality; neither of them, however, is defined in clear terms:

1. In a first sense, reality is composed by objects which, although not well defined, are similar to commonsense objects that we perceive through our sensory apparatus. The task of language is to *differentiate* reality into particular objects, and from this it follows that, whenever talking about reality, we always do it *relative to* language.
2. In a second sense, however, reality appears less determined and more confused, since it may be assimilated to a sort of *raw material* that produces the flux of experience or, to put it in another way, a sort of *substratum* that supports the sensible qualities of material things.

In both cases the notion of "ontology" is by no means clear. Should we adopt notion 1 or notion 2, or maybe both, but how? In Quine's view (or at least the first one, reported in the aforementioned quotation) ontological talk rises when the human mind—which cannot be clearly distinguished from language—gives an order to the disordered fragments of raw experience which, in turn, are given to us by the just-mentioned "substratum." This notion closely recalls a concept of "nondifferentiated reality" similar to the Kantian one. We know, however, that Quine rejects any distinction between a priori and a posteriori: he believes that any kind of ordering whatsoever cannot be separated by the act of theorizing (which, in his view, is tantamount to talking). In fact, he claims in "On What There Is":

> Our acceptance of an ontology is, I think, similar in principle to our acceptance of a scientific theory, say a system of physics: we adopt, at least insofar as we are reasonable, the simplest conceptual scheme into which the disordered fragments of raw experience can be fitted and arranged. Our ontology is determined once we have fixed upon the over-all conceptual scheme which is to accomodate science in the broadest sense; and the considerations which determine a reasonable construction of any part of that conceptual scheme, for example, the biological or the physical part, are not different in kind from the considerations which

> determine a reasonable construction of the whole. To whatever extent the adoption of any system of scientific theory may be said to be a matter of language, the same—but no more—may be said of the adoption of an ontology.[14]

The Quinean "ontological relativity," then, stems precisely from considerations of this sort. But it must not be forgotten that the roots of the problems we have met thus far lie in Quine's acceptance of Frege's views. According to Frege, existence is a property than can be expressed only in terms of logical quantification (and, more precisely, in terms of existential quantification). It is a property pertaining to concepts, and not to objects. Given this fact, the Fregean treatment of existence is a true turning point if compared to what the preceding philosophical traditions—and, in particular, the Aristotelian one—had claimed. It becomes in fact impossible—or, even better, *meaningless*—to talk of existence as if it were something belonging to such individual objects as Peter, Charles, Mary, and so forth. So we have a strong conflict with common sense (but Frege does not seem to worry much about this fact). This justifies—from a Fregean viewpoint, obviously—the claim that a sentence like *Michele Marsonet exists* is neither true nor false, but devoid of meaning. Despite many opposite and authoritative opinions, however, it seems to me that Frege's conception of existence gives rise to numerous objections at the philosophical level.[15] In any event, the endorsement of Frege's ideas has led the absolute majority of analytic philosophers to take more or less for granted the thesis that "existence is not a predicate," thus re-echoing the well-known Kantian statement (even though, it must be observed, analytic philosophers start from premises which are different from the Kantian ones). Since "exists" can never be compared to a predicate constant, its real meaning is detected by using the existential quantifier within predicate calculus.

Quine owes much to the Russellian theory of definite descriptions, while it is well known that Bertrand Russell is, in turn, indebted to the Fregean treatment of existence. It is worth mentioning, however, that the early Russell admitted various uses of the term "existence" that should be clearly distinguished:

i. according to the first use, which belongs to both philosophy and ordinary language, existence pertains especially to individual objects;

ii. the second use, instead, which has a more technical character, is employed in formal logic and is based on some features of mathematical reasoning.

Only later did the British philosopher fully endorse the Fregean paradigm, claiming that (ii) is the only possible use. But it is interesting to note that neither Russell nor Frege give convincing philosophical justifications for their choosing the logico-formal use of the term *existence*. We can only infer the presence of such justifications in a contextual manner, that is, by examining their works carefully. Even when such analysis is accomplished, however, many doubts still remain. It is therefore legitimate to ask the following—and, indeed, fundamental—question: "*Why* should we privilege the logico-formal meaning of the term *existence*, to the point of denying the presence itself of other—and equally legitimate—meanings of it?"

3. THE INDETERMINACY OF TRANSLATION

Many contemporary philosophers of science, and especially those who hold a relativistic stance, claim that there is just no external reality capable of explaining the truth—or falsity—of the sentences pronounced by members of linguistic communities that are more or less distant from us either in space or in time (or both). This means, in turn, to deny:

1. the possibility itself of any correspondence theory of truth; and
2. any kind of scientific progress.

But this view, however popular it is nowadays, may nonetheless be challenged. As is well known, in *Word and Object* and other subsequent works Quine claims that terms and their denotations are strictly linked to our grammatical analysis of the sentences belonging to our own language. So, when we have to translate a term used by a speaker of a completely unknown (to us) language—for example, the New Guinea aborigine previously mentioned—we have at our disposal an entire set of different choices, each of them complying with the dispositions to verbal behavior on the part of our foreign-language-speaker interlocutor.

Let us consider Quine's most celebrated example of the "gavagai." And let us also admit that, after all, the objects that my interlocutor means to name when he says "gavagai" are not rabbits, but rabbit's temporal slices or whatever kind of weird objects the fertile and endless phantasy of philosophers of language may invoke (someone, in this regard, wrote that many of the examples they make are "perverted"). Furthermore, let us forget the problems posed by philosophy of language and concentrate, instead, on the *practical* problems met by an empirical linguist working on a specific language of a primitive tribe (but not *exclusively* on language: I want to stress this fact, because he will surely take into serious account empirical and social factors as well). For—let us never forget it—empirical linguistics is a discipline which belongs to the realm of empirical sciences. Quine actually seems to think that an empirical linguist behaves like a formal logician, using variables, quantifiers, and so on. But this is not true, as any empirical linguist could easily confirm.

Day by day our linguist will become more and more acquainted with the language of the natives and, after a little while, having built up a reasonably good vocabulary, he will certainly be able to explain to them that, saying "rabbit," he refers to a particular class of animals, while when they say "gavagai" they refer to temporal parts of those same animals (assuming, of course, that a normal man—who is not a Quinean philosopher of language—may really think that way). It follows that the difference in our way of referring to some objects does not have the dramatic consequences foreseen by Quine and his followers. Differences may be easily settled by *practice*.

After all, this is the way real science works. If someone asks, Are you sure that those problems can be really overcome? I tend to answer, Give me an example of a primitive tribe whose language could not be adequately translated into our own Western languages. Indeed there is not even one, provided, of course, we do not have in mind a *perfect* translation, something no one has thus far been able to accomplish (and something no one will ever be able to accomplish either, just because perfection does not belong in our human world). And *this* is the point, because Quine has in mind the notion of "perfect translation."

Of course there are languages—such as Etruscan—that cannot be adequately translated into English (or Italian, or German), but this too depends on practical factors; that is, on the fact that today there are no Etruscans to talk with. If there were, I have no doubts that all problems concerning their language would be solved in a reasonably short time (and, obviously, we cannot rule out the possibility that, sooner or later, some kind of document or inscription may show up that will offer us the key to build a reasonably good manual of translation of Etruscan into English, Italian, German, etc.).

The solution to the problem is obviously much easier if we stick to more common languages. For instance, in Italian there is a word, *epistemologia*, which at first sight seems to correspond perfectly to the English term *epistemology*. It is a word belonging to the technical jargon of philosophy. If an Italian philosopher goes to Britain or the United States, however, he will soon realize that the Italian reference of the term is different from the English one. In Italian, in fact, *epistemologia* means "philosophy of science," so that, when I say "epistemologia delle scienze sociali," I mean, more or less, "philosophy of the social sciences." In English, instead, *epistemology* means "theory of knowledge," so that, if I say "epistemology of the social sciences," a British or American colleague—who does not know Italian—would not understand well what I mean when pronouncing that expression. Given these difficulties, we may even decide to use the two corresponding terms *gnoseology* and *gnoseologia* in order to translate "theory of knowledge," but this word sounds somehow outdated in both English and Italian. So it seems more reasonable to explain to my interlocutor the exact meaning of the Italian word *epistemologia*, giving him the opportunity to understand what I mean when using it, or we may even decide to forget about it, using only "philosophy of science" and "theory of knowledge" in our English conversation, just to prevent any misunderstanding.

In any event it is not a dramatic problem. It may be overcome with purely *practical* means: we bargain, as always happens in daily life. This leads me to stress what in my view is a very important point indeed. Quine—like most linguistically oriented philosophers—largely overestimates the role of *idealization* in philosophical work. True, we often need idealization both in philosophy *and* in science, but idealization is simply an *instrument* that is useful to clarify many issues; it must not become the final objective of our work.

Empirical linguists, for example, deal with *concrete* languages, and not with idealized languages (such as that of formal logic). As human beings we do not need a perfect translation of one language into another and, as a matter of fact, I believe that the notion of "perfect translation" does not even make sense: a reasonably good translation is sufficient for our purposes, the main reason being that of adequately understanding each other. An ideally perfect translation can never be achieved, due to at least two good reasons:

A. We are imperfect beings and language is not a kind of distant divinity totally separated from us or an a priori factor located in some sort of Platonic world. It is, rather, a set of social practices which, being created by mankind and being subject to *historical* vicissitudes, is imperfect, too, so even our attempts at translating one language into another are bound to be imperfect. But this certainly does not prevent a reasonably good communication.

B. Language is socially and historically determined. No one can prevent us from creating artificial languages, such as the one used in mathematical logic, that are supposedly perfect (although even this statement may be seriously questioned). But, in doing this, we always have to bear in mind that artificial languages are idealizations generated by our mind, and not something that can be found in reality itself. If we forget these facts, we are likely to build what I call a "paper ontology" which, beautiful and fascinating as it may be, is only a product of our mind and not something belonging to natural reality.

The logical positivists and many analytic philosophers (think of the great Polish logician Stanislaw Lesniewski, for example) have just taken this path, which is bound to be misleading, bringing us nowhere. And Quine is one of the champions of this enterprise, which becomes even more dangerous if we transfer it to the field of philosophy of science. There is, in fact, a good deal of the logical positivists' *philosophical utopia* in this way of doing philosophy.

Ian Hacking, for instance, makes a good point about the difference between natural and artificial languages. Speaking about reference and Hilary Putnam's growing anti-realism, he sets forth the following remarks:

> Putnam became sceptical. The malaise about meanings and Fregean senses owes much to W. V. O. Quine's doctrine of the

> indeterminacy of translation. Quine has a parallel thesis about reference: the inscrutability of reference. To put the idea crudely: you can never tell what someone else is talking about, nor does it matter too much. Quine asserted this with modest examples: where I speak of rabbits you might hear me as talking about spatio-temporal slices of rabbithood. Putnam adds real inscrutability. Whenever you talk of cats and mats, you might be referring to what I refer to when I speak of cherries and trees—yet the difference in reference would not come out, because anything I am confident of (some cat is on the mat) is expressed by a sentence which under your interpretation is something in which you have equal confidence (some cherry is on some tree). This is indeed extraordinary. . . . We need to have this bizarre claim made plausible to us.[16]

But surely this is a very demanding task, because it is difficult to make such an argument plausible for a normal person. In any event, it turned out that Putnam's bizarre conclusion was based on a famous theorem of mathematical logic: the Löwenheim-Skolem theorem. Putnam does nothing but make a generalization, applying it to any domain whatsoever of individuals, including cats, mats, cherries and trees. We may note, however, that here we have quite a big jump, since we are just applying a result concerning the artificial language of mathematical logic to the natural language we use in our daily conversation and, afterwards, to the technical language used by practicing scientists. And the jump is so enormous that Putnam's argument, although correct from a purely formal point of view, leads to paradoxical consequences. There is a Quinean influence here since, as everybody knows, Quine says that scientific sentences must be translated into first order logic because in his opinion (i) this move will guarantee the perspicuity of scientific talk, and (ii) it will render the hidden ontological committments of scientific theories perspicuous as well.

This proposal, however, does not seem to be very popular with operative scientists who, after all, can do their job even while ignoring formal logic. But the most important thing to be noted is that Putnam's application of mathematical logic's results to ordinary and scientific languages—and to the language used in daily life as well—is likely to create more problems than it was meant to solve. This happens just because the three languages have different origins, different stories and different purposes. Every time you try to "squeeze" the commonplace language of physicists or ordinary language spoken in daily life into first order logic, your move will not work because they—the three different languages—will not fit. Adopting a terminology borrowed by the later Wittgenstein, and thus remaining within the field of philosophy of language itself, in my view Putnam does not realize that different *language games* are here involved, so that playing the language game of physics using first order logic will surely cause troubles to us, and the same can be said about the common language used in everyday life. It is worth noting, too, that this is even more weird if we keep in mind that Putnam is very fond of the later Wittgenstein: he always declares his intellectual debt to such works as the *Philosophical Investigations* and *On Certainty.* But no doubt the main shortcoming of Putnam's analysis is due to the fact that he binds himself to purely linguistic considerations, thus not taking into account the practical and operational aspects of both empirical science and daily life. So, as I said before, the philosophical utopia of logical empiricism is still at work here.[17]

Going back to Quine once again, we may recall that he, especially in the 1950s, followed quite a different path. In "Two Dogmas of Empiricism," in fact, he explicitly recognized the soundness of a *pragmatist* vision of the working of language and science. Later on, however, his ties with pragmatism became looser. Perhaps we can identify the reason for his change of mind in the fact that Dewey and the other main figures of American pragmatism always stress the *practical* side of both linguistic and scientific enterprise, thus not giving too great an importance to the creation of artificial languages. Probably this is a fault in Quine's eyes, while it may be a merit from a different point of view.

In any event, this may also explain why Quine and many analytic philosophers so often lose sight of reality in order to chase their logical (formalistic) dreams. The pragmatists adopt, as we remarked in the first chapter of this book, quite a different stance, and Quine had to choose: his choice eventually favored a one-sided vision of formal logic and logical analysis of language. Just for this reason, as I said previously, he is still very close to logical positivism, despite the deep and refreshing insights contained in "Two Dogmas of Empiricism." No doubt things would have taken a different course had Quine pursued in the later years his pragmatist inclinations. It is certainly correct to say that he deserves the merit of having underlined the importance of pragmatism in the United States during a period which saw a large predominance of logical positivism, but it is correct as well to claim that later on he somehow betrayed this promising insight of his.

On the basis of what we have said thus far, there are indeed good reasons for claiming that philosophers of science are not supposed to build their own theories independently of scientific practice proper. If they do that, their theories may certainly be fascinating from a purely philosophical and logical viewpoint but, to say the least, those theories will turn out to be useless for practicing scientists (who, after all, should be the ones who provide the philosopher *of science* with the raw material to work on).

As I briefly hinted in the first chapter, the prevalence of a linguistically oriented philosophy of science explains why, nowadays, most scientists do not seem to take contemporary philosophy of science seriously. While it is correct to say that scientists took seriously the theses of logical empiricism (although often criticizing them) and Popper's ideas, they look embarrassed when reading either Feyerabend's claim that man's landing on the Moon has no objective value and that scientists, in principle, do not know reality better than a shaman of any primitive tribe, or the cat/mat/cherry/tree story narrated by Putnam. Lately philosophy of science tends to grow as a purely philosophical discipline. But, destroying the essential bridges with science as it is really practiced, it becomes a merely *speculative* enterprise whose right to be called "philosophy of science" is questionable. This is a great loss indeed, for both philosophers and scientists because, on the one hand, philosophers need healthy contacts with science just to avoid their perennial tendency to talk about things that exist only in their minds, while on the other hand, scientists, being obviously very much involved in the practical problems of their daily technical work, usually do not have time to take into account the philosophical problems posed by scientific development (but still need to be alerted about these same philosophical problems).

Someone might raise the following objection to what I have thus far claimed: After all, we can overcome the pitfalls due to a purely linguistic approach to philosophy of science remaining *in* the field of philosophy of language. It is possible, in fact, to carry on as many critiques as we wish on Quine's account using the classical tools of philosophy of language and formal logic[18] and, given this fact, it follows that we do not need to abandon philosophy of language altogether. For example, we may defend a realist account of the progress of science using theses not about meaning but about *reference*, showing that competing or successive scientific theories are usually about the same things.[19] To this objection, which has some good foundations, I reply in two ways.

First of all, it is obvious that my criticisms of the philosophy of language do not apply to the discipline as such, but simply to the bad use that many philosophers of science today make of it: philosophy of language is indeed necessary. Second, it is true that philosophy of language may have such a role, but we must be careful. I have no objections to using linguistic analysis—within the field of philosophy of science—as a *tool* to obtain nonlinguistic results. But, on the other hand, we should realize that the problems posed by the foundation of scientific realism are not only linguistic but also—some would say mainly—practical and operational problems. Obviously linguistic analysis is a precious instrument which, being now at our disposal following the impressive developments of analytic philosophy of language in our century, may be adequately used in order to clarify the nature of scientific questions. But I insist on my caveat: it is dangerous to assume that philosophy of language can solve those problems. It simply cannot do that due to the nonlinguistic character of the problems themselves. And much the same can be said about formal logic: it has a precious role within philosophy of science as long as we understand that, in this context, its role is most of the time instrumental and auxiliary. But when it comes to philosophy of science, logical and linguistic analyses are not sufficient, because they are not meant to take into account the pragmatic dimension of scientific enterprise (and precisely *that* dimension is essential in science).

4. NEURATH'S CONCEPTUAL SHIP

As I remarked before,[20] I deem it correct to claim that the famed metaphor of the "conceptual ship" introduced by Otto Neurath plays an absolutely fundamental role in Quine's thought. Let us see what the American philosopher claims in an essay which first appeared in 1950 in the *Journal of Philosophy*:

> The fundamental-seeming philosophical question, How much of our science is merely contributed by language and how much is a genuine reflection of reality? is perhaps a spurious question which itself arises wholly from a certain type of language. Cer-

> tainly we are in a predicament if we try to answer the question; for to answer the question we must talk about the world as well as about language, and to talk about the world we must already impose upon the world some conceptual scheme peculiar to our own special language. Yet we must not leap to the fatalistic conclusion that we are stuck with the conceptual scheme that we grew up in. We can change it bit by bit, plank by plank, though meanwhile there is nothing to carry us along but the evolving conceptual scheme itself. The philosopher's task was well compared by Neurath to that of a mariner who must rebuild his ship on the open sea. We can improve our conceptual scheme, our philosophy, bit by bit while continuing to depend on it for support; but we cannot detach ourselves from it and compare it objectively with an unconceptualized reality.[21]

Furthermore, it is worth stressing that Neurath's metaphor plays a central role not only in Quine's thought but in the entire speculation of the logical positivists and of most analytic philosophers as well. There is a clear Kantian heritage here: we can never know reality-in-itself (i.e., more or less, the *noumena*), but can nonetheless know reality as we *say* it is (i.e., the *phenomena*). This sort of Kantian dualism is precisely the source of many of the problems with which Quine, logical positivists and analytic thinkers cope. Because if we assume that we can never reach reality-in-itself, it is obvious that our conceptual schemes are only bound to determine what reality is *for us*, and it is clear, too, that the language we speak becomes something much more important than a simple instrument for *referring to* a reality which is nonlinguistic in character. The problem is twofold:

1. Do we really need to assume that there is an unbridgeable gap between reality as it is and reality as we know and talk about it, and
2. how justified is it to claim that we can never reach reality-in-itself?

Obviously, priority must be given to question 1, since question 2 disappears as long as we stop answering in the affirmative to question 1. Instead of taking a purely philosophical path getting involved, for instance, in the current disputes between realists and anti-realists,[22] I would like to stay close to what some scientists actually do, reconsidering the examples concerning paleontology, astronomy, and cosmology that I already used in the first section of this chapter.

In the last century, paleontologists made many discoveries about the early ages of the Earth, when no human beings were present on the surface of our planet. They found lots of evidence regarding such strange living beings as lizards with feathers who were able to fly, or enormous reptiles that are today called dinosaurs. It is indeed difficult to deny the past existence of these living beings, because their fossil prints have been found in many places throughout the five continents. From a philosophical viewpoint one may say that, although no longer existing, they were part of our world in those far distant eras. Ontology can thus be modified, but changes are due to the modifications which took place in reality itself: they are *not* a consequence of modifications that took place in our—human—conceptual schemes.

Now, let us ask ourselves the following question: What does it mean, in the case of these discoveries of paleontology, to claim that we cannot reach reality itself, but just reality as we know or talk about it? Perhaps the following tentative answers may be given:

1. The fossil prints do not offer sufficient evidence for the existence of the above-mentioned living beings. Maybe due to some strange chemical reactions, the stones have assumed that form and therefore we erroneously think that they were made by animals living in the past.
2. It is also possible to assume that a mean daimon disseminated fossil prints on the Earth just to deceive us. There were no temporally distant eras on our planet, but just an endless present in which we are immersed without any chance of understanding why things are the way they are.

Note that arguments quite similar to (1) and (2) have been often set forth throughout the history of philosophy, but usually practicing scientists do not take them seriously. At this point, someone should explain what it means to claim that dinosaurs exist only in connection with my conceptual scheme, given that paleontology gives us overwhelming evidence that dinosaurs were thriving on Earth before mankind's appearance on our planet and, consequently, before any human conceptual scheme began to work. Unless the anti-realist means to claim that our conceptual schemes (or "scheme" if he believes there is just one that covers all human beings' minds) are time- and space-independent (i.e., they exist from eternity). But this is unwarranted metaphysics, and I do not think that a serious anti-realist wants to make such an assumption. The only real way out at our disposal is rejecting what science—in this case paleontology—says on the basis of empirical data. But, obviously, this way out leads immediately to irrationalism.

In any event, the fact of the matter is that our conceptual schemes play no essential role in this case. We are simply talking about forms of life which were present on the Earth *before* mankind, and this offers sufficient ground (although not a logically conclusive one) for claiming that there is a reality in itself, independent of any way men conceptualize and judge it. And much the same can be said when dealing with modern cosmology. When astronomers and cosmologists succeed in spotting galaxies whose age largely exceeds that of our solar system, that is enough to abandon any anthropocentric vision of reality, be it the Kantian one—the famous "Copernican revolution"—or the vision favored by many linguistically oriented philosophers (including those who claim they are philosophers of science).

Note, too, that this fact is likely to have an enormous impact on *ontology.* Quine, for instance, claims that our ontology is always relative to the language we speak, so that the adoption of an ontology is a largely arbitrary operation. From this follows, among other things, that there is no absolute ontology, but as many ontologies as we like. But this is simply false: Quine blurs the distinction between the world on the one side and what we think *about* the world (or what we say *about* it) on the other.

When someone claims, since we have not yet been able to ascertain the structure of most microphysical particles, that those particles are unknowable, he just forgets to take into account the *not yet* which plays a crucial role here; history of science is crowded with supposedly unknowable entities which, later on, were found to be perfectly knowable. And this is exactly the basis of that scientific progress so many philosophers of science today vigorously deny. So when Paul Feyerabend—just to quote a very popular philosopher of science of our days—claims that there are many ways of judging reality, one could answer that this is certainly true, but with a very important caveat. History of mankind shows that we can both judge and talk about reality in many ways. However, as I pointed out previously,[23] when we question nature we receive different answers, and nature itself shows that some of our questions are legitimate while others—many others—are *not*. And this simply means that reality does not react in the same way to our attempts to get to know it. It is in fact reality itself that tells us whether our attempts do—or do not—go in the right direction. And, furthermore, this is another step toward recognizing that language cannot be self-sustaining: since it is essentially referential, it always needs a second factor (that is, reality) with which to match.

This is precisely the heart of scientific progress. Men abandon their attempts which were not successful and pursue instead the ones which were, trying to get better and better results and trying to get as close as possible to the structure of reality itself as well. Again, it is difficult to see in this case why reality should exclusively depend on our conceptual schemes.

5. LINGUISTIC IDEALISM

Even analytic philosophers and, in general, linguistically oriented thinkers must face the Kantian heritage. The problem in this case, as I hinted before, is that most of them depict themselves as empiricists or realists, while a careful scrutiny of their theses clearly shows that they are what I call "linguistic idealists." I would like to make clear, however, that I do not attach an intrinsically negative meaning to the term *idealism* (although I do not find it congenial to my convictions); such a move, besides being philosophically improper, is bound to disregard the important fact that idealism is a philosophical doctrine that can be reasonably argued for. My claim is, instead, that no one should be allowed to declare officially that he is an empiricist, while his theses show that he is actually giving an idealistic picture of reality. Unfortunately, such is the case with many important analytic thinkers nowadays, and I have already argued in this work that this is true even of an author like W. V. O. Quine, who is commonly regarded as the champion of empiricism within today's philosophy.

This fact can be better understood if we remember that both logical positivism and the analytic tradition have always been tremendously ambiguous on the real nature of the relationship between *language* on the one side and the *world* (reality) on the other. The standard interpretation of logical positivism and of the analytic tradition claims that linguistic analysis is the best way we have at our disposal for getting rid of both metaphysics and any nonempiricist interpretation of reality. *If* it were so, it would be correct to say that linguistic analysis is the most effective anti-idealistic method devised by twentieth-century philosophy. But this picture, although still widespread and successful nowadays, is both inaccurate and misleading. For it does not take into account the essential fact that the linguistic turn, despite the strong empiricist and antimetaphysical commitments of its founding fathers, has increasingly viewed language as an a priori element which, in turn, determines and even *builds up* reality. Many analytic thinkers—especially in the last few decades—have taken an even bolder stance, proclaiming fearlessly that reality *is* language, so that, for instance, it is our conceptual schemes that determine reality (and not vice versa).

At this point the shadow of Immanuel Kant comes on the stage once again. According to the main representatives of logical positivism, the philosopher's task is to find out what the nature and the extension of any meaningful discourse are and, in so doing, set the limits of scientific inquiry itself. As I said before, speaking of the metaphysical tenets of logical positivism,[24] the philosopher thus becomes a sort of "superscientist" who determines the bounds of sense, and practicing scientists must rely on his advice to find out whether their work is meaningful or not. Linguistic analysis, therefore, becomes a sort of *prima philosophia*, that is to say a superdiscipline that determines the conditions which regulate our knowledge. But we must also recall that, according to Kant, no pure perception of reality is possible without the mediation of our ability to conceptualize; the categories are always needed in order to know the world. It follows that no absolute knowledge of reality is ever possible, the only knowledge at our disposal being the one relative to the conceptual apparatus. And while Kant believed that the conceptual apparatus filtering experience is located in human intellect, the analytically oriented thinkers locate it in language. Both, in sum, speak of *preconditions* regulating human knowledge, be they a priori categories or language. In both cases, it is philosophy that answers the basic questions, and this fact gives further ground to the thesis I previously stated: the antimetaphysical stance of logical positivism and of the analytic tradition is more theoretical than real. It looks more like a philosophical slogan than an effective accomplishment.

Just as Kant thought that our knowledge of the world makes sense only with reference to a system of a priori categories, so the analytically oriented philosophers deem it correct to say that our discourse about the world may be meaningful only as long as it is referred to a system of linguistic representation. But what is "language" if we start from these premises? Certainly it is neither a part of the world nor a tool devised by mankind for accomplishing communicative—i.e., social—purposes. Language becomes, rather, an a priori factor which categorizes reality. It makes no sense to ask where it comes from, because its connection to reality is something inexplicable. And, taking that path, sooner or later one is bound to conclude that language and reality are exactly the same thing, a position that many representatives of the analytic tradition are actually endorsing in the present days. They are simply developing the Kantian position in a consistent manner—as the classical idealists did in the nineteenth century—shifting the attention from epistemology to philosophy of language and the theory of meaning. All this is confirmed by the following—and very interesting—remarks put forward by David Pears:

> Wittgenstein's leading idea was that we can see further than we can say. We can see all the way to the edge of language, but the most distant things that we see cannot be expressed in sentences because they are the preconditions of saying anything.[25]

Note how close this notion of "preconditions" is to the Kantian one. We may wonder, however, why it is necessary to suppose that there is such an unbridgeable gap between what can be said and what can be seen. And, furthermore, why should we assume the notion of the "edge of language" to be real and correct? It seems to me that many of the analytic tradition's basic theses rest, once again, on a notion of language conceived of as a supernatural (i.e., a priori) factor.

If we reject this notion, sense-conditions are no longer "ineffable," because they must be traced back to reality conceived of as a whole of which man is just a small part. And this means that the doctrine of the *limits of language* is likely to take us up a blind alley (although it is fascinating from a purely philosophical viewpoint). So, when Wittgenstein claims

> Objects can only be named. Signs are their representatives. I can only speak about them: I cannot put them into words. Propositions can only say how things are, not what they are. The requirement that simple signs be possible is the requirement that sense be determinate.[26]

We are entitled to ask, *Why* should we want "to put the objects into words"? And why is it necessary to assume that a sort of "veil" hides what things really are,[27] thus presupposing the existence of some mysterious essence of things that our language as such is unable to capture? It seems to me that in this case—as in many others—the philosopher (i) first invents the problem, and (ii) later on claims that it cannot be solved. But we must solve *real* problems, not artificial ones. All these difficulties stem from the refusal to consider language as a part of nature, that is, as a natural tool.

CHAPTER 3

Realism and Anti-Realism

1. SOME CURRENT ANTI-REALIST OBJECTIONS

Let us now review some of the main challenges anti-realism poses to a realist view of scientific enterprise. I will not examine in detail what particular anti-realist thinkers today say, limiting myself with taking into account some general arguments in order to show that anti-realism can indeed be reduced to a few basic theses (no matter the form in which they are actually put forward). In the *Critique of Pure Reason,* Kant claims that the world in itself is unknowable. He does not deny that there is a reality independent of our knowledge, which may after all be the basis of knowledge itself. What he rejects, instead, is our capacity to describe and know this purported independent reality. This situation is due, in his view, to the fact that in order to express our judgments we use a priori concepts and categories, which have nothing to do with an alleged external reality. Kant's transcendental idealism, thus, says that our judgments are universal and necessary but, on the other hand, they can never reach reality as such. Now, we know that these original Kantian theses are questionable today, given modern science's results. For example, we may try to draw a distinction between existence, which is absolute (as Quine himself sometimes recognizes[1]), and our knowledge of it, which is not. Like the American continent before Columbus discovered it, so reality exists independently of our experiencing it or not (and this is true even when we get to know a preexisting reality only *afterwards*). Kant, for instance, claims:

> The objects of experience, then, are *never* given *in themselves*, but only in experience, and have no existence outside it. That there may be inhabitants in the Moon, although no one has ever perceived them, must certainly be admitted. This, however, only means that in the possible advance of experience we may encounter them. For everything is real which stands in connection with a perception in accordance with the laws of empirical

> advance. They are therefore real if they stand in an empirical connection with my actual consciousness, although they are not for that reason real in themselves, that is, outside this advance of experience. Nothing is really given us save perception and the empirical advance from this to other possible perceptions. For the appearances, as mere representations, are in themselves real only in perception, which perception is in fact nothing but the reality of an empirical representation, that is, appearance. To call an appearance a real thing prior to our perceiving it, either means that in the advance of experience we must meet with such a perception, or it means nothing at all.[2]

Once again, we may note how close many contemporary analytic philosophers are to Kant's positions. We can answer this way. *If* the inhabitants of the Moon are real, they are so independently of the fact that we *think* that they are or *talk about* their existence. And much the same can be said if they are not real. Our thought and language—or the language/thought that is the element on which the speculation of many linguistically oriented philosophers is based—does not modify the ontological status of reality. If later on we find out that there really are inhabitants on the Moon, our discovery does not make any difference to their *real* existence. They were real before our discovery and will continue to be real following it. The fact that on October 12, 1492, Christopher Columbus discovered America certainly modified human history, but adds nothing to the ontological status of the continent that was named America and existed already before Columbus's discovery. And—of course—if someone denies its existence prior to Columbus's voyage I really do not know what to answer: any rational discussion must end at this point.

As far as science is concerned, after all, I think it is easy to verify that it is interested in a reality which is independent of our knowledge: this is true not only for paleontologists and astronomers but for physicists, biologists, and others as well (even though this is no final argument against anti-realism). As we just said, today we can challenge some of Kant's central theses thanks to the discoveries made by modern science. In particular, scientists found out that Euclidean space and absolute time are not a priori. Relativity theory shows that physical space does not have an Euclidean character, and this is not an idea imposed upon reality by our conceptual schemes but rather a discovery about a world whose structure does not depend on our knowledge of it. Much the same can be said about Kant's categories; quantum mechanics, for instance, showed that causality is not a rule that has no exceptions. But the idea that the world in itself is unknowable is still very successful nowadays. It has in fact been reproposed by many philosophers who want to stress the role of subjective factors within knowledge.

A further and important argument is the following: We do not know reality *directly*, but only through representations such as ideas and mental images. If this is true, it follows that we only know our representations, while it is impossible to know an alleged reality in itself. However it should be easy to realize that, from the fact that we know *through* representations, it does not follow that we can only know representations and nothing else (what about the *causes* of representations, since it would be hard to claim that representations come from nothing?). We might even take into account the necessity of subjective and a priori elements in knowledge. Psychology, in fact, purportedly shows that our conceptual apparatuses vary a lot depending on time, civilization, scientific world-views, and so on. If so, conceptualization becomes a sort of veil which prevents us from directly reaching reality, and all this in turn means that knowing always implies judging and applying concepts. These are the bases that led many twentieth-century philosophers to criticize the so-called myth of the given; nothing of what we passively get from reality is by itself an epistemic content. The presence of this "veil of conceptualization," however, may lead us in different directions. Someone may claim that it is not possible to compare our image of reality, which has a conceptual character, with a reality which purportedly is nonconceptual or preconceptual. But this alleged impossibility implies just a theoretical risk: that is, the fact that we do not actually know the alleged world as such, and not the final impossibility of knowing it.

Yet another argument says that our knowledge does not concern an independent reality because is not based on data referring to that reality. In fact, the "foundations" of knowledge (protocols, etc.) always describe a conceptualized world. This means, more or less, that all our information is already conceptualized. We can note, however, that by adopting this style of reasoning we lose the totality of the picture. *Where*, in fact, does our information come *from*? Certainly we do not create it independently of something which is external to us, unless we feel inclined to admit that the world is a mere product of our mind. But there is even more to say. Let us accept, for instance, that no preconceptual datum—a sensory stimulus, for example—is by itself an epistemic content. On the other hand, it is certainly reasonable to claim that the preconceptual data *do* make a decisive contribution to epistemic contents. It is therefore incorrect to conclude that, within knowledge, there is nothing that is given to us by external reality. It follows, of course, that the data coming from an external reality which is independent of our knowledge have a crucially important role in determining our knowledge, although it can easily be admitted that those data do not completely determine the form of our knowledge. All this means, eventually, that there is effectively a reality which does not need our conceptualization or categorization in order to be (exist).

A further argument against the knowability of the world claims that, even granting that our knowledge is really based on some preconceptual inputs, we must nevertheless admit that, in order to perceive, think, or describe something, we do so in conceptual terms. But, then, we will never be able to answer the question how the world in itself really is, because any kind of reply is already based on concepts. And this, more or less, is tantamount to claiming that any time we try to speak about an objective and mind-independent reality using "mentalistic" terms, we are likely to run into contradiction. Such an argument is based, however, on questionable assumptions. In other words, it implicitly assumes that knowledge is some kind of *copy*, that is, an object that exactly reproduces a second object which is known and has all—and only—its characters. And since our representations lack at least one, very important character of the alleged world in itself (i.e. *nonconceptualization*), or have at least one, very important character that the world in itself does not possess (i.e. *conceptualization*), it follows that we cannot know it.

Why is this argument based on questionable assumptions? For at least two reasons: first of all, knowledge is neither a representation nor an object; secondly, representation does not need to be a copy (nor can it be). A representation never reproduces all characters of the object it is meant to represent: this is true of a photograph and of a linguistic expression as well. In a linguistic representation—as Wittgenstein showed—the representing object and the represented one have no concrete characteristic in common, but just some *structural* proprieties. So we may say that in order for us to know an object, a representation must share with it only *some* of its characteristics, which are reproductions of similar features of the object that is going to be known. As Wilfrid Sellars would have said, knowledge is no replica or reproduction, but a sort of projective relation between the structure of reality on the one side and mental states and practical activities on the other. And things must be that way because otherwise material reality would be totally unknowable to a conscience whose main job is conceptual translation, especially with regard to physical events. Thus it is incorrect to claim that conceptualization prevents our knowledge of an external world: the fact that we talk and describe it in conceptual terms by no means implies that conceptualization regards reality as such, too.

Note—and this is really important for the philosophers of science—that the correspondence between reality and our representations of it is able to explain why, after all, science is successful (I mean *practically* successful: scientists get things *done*). So, it is true that any sentence we pronounce depends on a particular conceptual scheme or reference system, but this does not compel us to deny that it depends, too, on the way reality is made independently of our judgments. There is always an involvement of two variables as long as our knowledge is concerned (reality is certainly "for us," but there must be a reality *as such* before there is a reality for us). Let us never forget, furthermore, that science gives us insights into a reality without mankind. It is interesting to note that Hilary Putnam, in a book made out of three lectures he delivered at the University of Rome and recently published in Italy, replies to some arguments by Hacking in the following way, commenting on Hacking's famous statement concerning positrons, "If you can spray them, then they are real":[3]

> But what does it mean to believe that "they" are real? If it means to believe that *distinct things* called "positrons" do exist, then we will have problems—and serious problems indeed—with the theory. . . . I imagine, at any rate, that he [Hacking] wants to tell us that *here*, in *this* experiment, positrons are "real" *without however telling us what that means*. . . . Naturally I do not intend to say that positrons are not real. But our believing in their reality has a conceptual content only because we have a conceptual scheme—a very peculiar scheme, a scheme we do not fully understand but that, nevertheless, is successful—which makes us know what must be said about positrons. Hacking's

> attempt to create a clear distinction between facts and theories . . . collapses because of the interrelations between facts and theories . . . the term "positron" is no *copy* of a reality but rather a "symbol", and it is *theory* that teaches us how the symbol should be used.[4]

Without repeating things I have already said before, let me only stress that, in adopting this stance, Putnam makes scientific objects depend on human conceptual schemes, and this is an unavoidable consequence if we claim that language (or maybe it would be better to say, once again, "language/thought") has a sort of ontological priority over reality. This is just a first step because, usually, those who adopt this position are bound to go even further, claiming that reality and language are exactly the same thing. I would like to note that, in the experiment named by Ian Hacking, the scientist does after all *do something* in order to obtain a certain—practical—result. But doing something implies interacting with *something else*, and this something else cannot certainly be our conceptual scheme, unless we want to grant that men create reality.[5] This means that the conceptual scheme, however important it is, must take into account a *something* that is different from it: our theorizing and conceptualizing could not even occur if they did not start from the necessary inputs which come from a mind-independent reality. Furthermore, it may be true that the theory tells us what to do with positrons, but does it follow that the reference of positrons can be identified with the theory?

The importance of our previous short remarks against Kant's view of space and time should by now be clear. Since it has been shown—by contemporary science and by philosophers like Hans Reichenbach as well—that space and time are not characteristics that our conceptual apparatus imposes upon reality but, rather, features of reality itself, we no longer need to assume that a hypothetical inhabitant of the Moon is real only as long as he "stands in connection with a perception in accordance with the laws of empirical advance," as the philosopher of Königsberg tells us. We must understand, in fact, that this enormously successful—within philosophy, at least—Kantian stance is just the opposite of what current scientific practice teaches us.

If we ask an astrophysicist whether he believes that other planets of our galaxy are inhabited, he will most likely give us a prudent and interlocutory answer of the following kind: "Actually we have no evidence supporting this hypothesis. I can only say that we can neither confirm nor rule it out." What is clear, however, is that our conceptual apparatus has nothing to do with the existence (or nonexistence) of any extraterrestrial intelligent creatures. If they are somewhere in our galaxy right now they are real; otherwise, they are not. It is simply a *contingent* matter that, in this moment, we are not able to give a precise answer to the question. Furthermore, it is precisely the fact that our present scientific instruments cannot overcome the strictures imposed on us by space and time conceived of as characteristics of physical reality that prevents us from giving a precise answer. In any event, we have no reason at all to assume that their existence depends on alleged a priori categories which would allow human beings to build up reality.

We can of course ask ourselves the following question: Is there an internal time, neatly distinguishable from physical time, and such that it would authorize us to maintain a weaker Kantian assumption? However popular this position may be nowadays, I do not see any good ground for accepting it. Consider two persons—A and B—who are waiting for a bus. A is in a hurry because she has a staff meeting in less than half an hour, and so time seems to her to pass too slowly. B instead has an appointment with a person he does not like at all, and so time seems to him to pass too quickly. Here we have two very different perceptions of time. No doubt, however, there is an *objective* time which is measured by both A's and B's wrist-watches, and that is not affected at all by A's and B's personal perceptions. That is exactly time as a characteristic of physical reality that is, as relativity theory teaches, inseparable from space: men do not create time at all. And we should indeed be very careful when hearing arguments about the creation of time by human subjects, because those who advance them usually use them as premises for going even further. For instance, it is rather frequent nowadays to meet philosophers of science who claim that, since our ancestors held a different picture of the world, then they lived in a different world altogether. But this is clearly a non sequitur. Although the Ptolemaic picture of the universe was different from ours, still neither the Ptolemaic vision nor Newton's or Einstein's have the power to change the way reality is.

The question of the relationship between the knowing subject and reality calls for brief consideration at this point. For reasons that should by now be clear (given what we have said thus far), I will not adopt the Kantian expression "reality in itself." By the term *reality* I simply mean reality as long as it is independent of any human being's mind/thought/language. Let us again ask ourselves the fundamental question: Do we have serious reasons for denying that reality is knowable? Someone might answer that to know necessarily means having some kind of intercourse with what is known. It follows that, since we can only get to know previously unknown objects (entities), it is never the case that reality is independent of our knowledge of it. Here Kant's influence is still at work. Those who deem this argument important and conclusive do not realize that they are blurring the distinction between reality and our knowledge of it. When scientists look for features of reality which are still unknown to mankind—think, for example, of all the researches currently concerning the subatomic particles—they construct models which are supposed to represent reality. Subsequently, some models are retained and others are rejected. But this by no means implies that there is no reality independent of the knowledge we may have of it: the continuous strife towards new scientific models shows that just the opposite is true. As I said before, the fact that we do not have definitive answers at hand has a contingent character, and not a necessary one. In any event, we may also note that it is correct to maintain that, once we get to know reality, this same reality has some type of special relation with us, the knowing subjects. But it is unjustified to infer from this that the ontological status of reality depends on our knowledge. The example previously made of Christopher Columbus's discovery of America[6] may well illustrate this point. Certainly his trip made the continent which was later named America known to the Western world, but it would be absurd to claim that this same continent had no ontological status prior to 1492. The fact of being known—or unknown—to us is just a contingent fact which, as such, does not affect the ontological status (i.e., the existence or the nonexistence) of any particular portion of reality.

2. FURTHER OBJECTIONS

Some important objections may be raised to what I have said thus far. For example, against my deploying the idea of a person-devoid (and thus language-devoid) possible world, one may object that we must distinguish between what is true *in* a world and what is true *of* it.[7] In fact, even that person-devoid (and thus language-devoid) world is involved with language-using persons: namely, ourselves who suppose or imagine it. Quine, for instance, might contend that the ontological structure of that world is obtained through language-users, that is, we who specify what possible world is at issue.

First of all, let me remark, as I did previously, that the expression "possible world" is not appropriate in this context, because it resembles too closely the notion currently utilized in modal logic and analytic metaphysics. To me, the worlds with which paleontology and cosmology are concerned are just as *real* as our actual world. There is no mere "possibility" there, in the sense that, thanks to scientific research and instruments, we get overwhelming evidence about their concrete existence in the past (paleontology) and even in present and future times (astronomy and cosmology). What I want to make clear, in sum, is that the dinosaur whose skeleton I saw yesterday at the Natural History Museum was *really* thriving on a *real* Earth when no man (and thus no conceptual scheme) was around, and the proof is given by the fact that a part of its body—the skeleton—is still here. Likewise, the photographs taken by the Hubble Space Telescope show that I am talking of real objects, and not of "possible" ones. We can fairly accept the fact that, in order to talk about those objects or worlds, we need language (and that, furthermore, to think about them we need thought). Note, however, that the impressive consequences Quine and many other linguistically oriented philosophers draw from this fact are no longer there if we drop the Kantian conception of time and space as a priori requisites for organizing reality and adopt, instead, a more naturalistic (or—better—more scientific) manner of conceiving time and space.

Of course one could ask, Why should I do that? The answer is not difficult. Kant, in fact, idealized Euclidean geometry and Newtonian physics, believing that they offered, once and for all, *the* right picture of reality. The problem is that contemporary science shows instead that Kant was wrong, even though a statement like this is bound to cause waves of bewildered indignation among most philosophers. As I said before, today we can challenge some of Kant's central theses thanks to the discoveries made by modern science. This explains quite well why the physicist Steven Weinberg, when talking of Kant's influence in current philosophy of science, argues this way:

> Although naive mechanism seems safely dead, physics continues to be troubled by other metaphysical presuppositions, particularly those having to do with space and time. Duration in time is the only thing we can measure (however imperfectly) by thought alone, with no input from our senses, so it is natural to imagine that we can learn something about the dimension of time by pure reason. Kant taught that space and time are not part of external reality but are rather preexisting structures in our minds that allow us to relate objects and events. To a Kantian the most shocking thing about Einstein's theories was that they demoted space and time to the status of ordinary aspects of the physical universe, aspects that could be affected by motion (in special relativity) or gravitation (in general relativity). Even now, almost a century after the advent of special relativity, some physicists still think that there are things that can be said about space and time on the basis of pure thought.[8]

But if space and time are no longer a priori factors that allow me—the knowing subject—to give an order to the world, but rather ordinary characteristics of physical reality, I have no reason to think that I (along with my conceptual scheme) deserve a very special and unique place in the system of reality itself; this means, in fact, that I do not determine the way reality is, in the sense that time and space are quite independent of my (and any other human being's) conceptual scheme. Note also that, in taking this path, I no longer need to assume that things are "real" insofar as they get in touch with my perceptual apparatus, as Immanuel Kant claimed. Neither need I assume any longer that language is an a priori factor (as thought was in the speculation of classical idealism).

At this point we may also concede that there is a difference between what is true *in* a world and what is true *of* it. Obviously, it would be much better if we were able to travel back in time and see just how the Earth was before mankind's appearance, rather than simply talking about it: the fact of being *in* that world would make our judgements much safer. But if we give up the idea of space and time as a priori factors (and science offers good grounds for doing that), and if, furthermore, we give up the idea of *perfect* knowledge, then the difference between what is true *in* a world and *of* it becomes far less important. Surely we get from science—

think again of paleontology—incomplete information, but its amount and extent is such that I can have confidence in two facts: (i) there really was a time in which no man (and therefore no conceptual scheme) was present on our planet; and (ii) having dropped the important Kantian assumption mentioned above, the existence of that world no longer depends, so to speak, on my getting in touch with it through perception. Man, his language and conceptual scheme become, along with space and time, *natural* components of reality.

The second objection starts from the assertion that we need recourse to a reality that is totally independent of language and thought, and says that someone might observe that we cannot say *what* this is like. So we must distinguish between the *that* and the *what* of this mind/thought independent reality, and can sensibly claim *that* it exists while simply rejecting the challenge to specify *what* it is like. Just to make an analogy, I know *that* there are errors in present-day science, but cannot say *what* they are. But the objection is valid only as long as I claim that we have a perfect knowledge of that reality, something I never said because—as it was frequently pointed out in the preceding sections of the present work—I do not think that the category of "perfection" belongs in this world. I would never dare to say that our actual science gives us the true and perfect picture of reality (but Kant *did* make exactly that mistake with regard to Euclidean geometry and Newtonian physics). So I do not assume at all that Einstein's relativity theory gives us *the* true picture of reality, since I know well that, in a future we cannot actually foresee, it will probably be replaced by a better and more comprehensive scientific theory. What I claim, instead, is that today Einstein's theory is undoubtedly better than Newton's; therefore, it is absurd to adopt a sort of Kantian dualism, which is based on the overevaluation of two scientific theories that have subsequently been shown to be valid only for a limited portion of reality. But Rescher has some reasons for claiming that

> The current state of "scientific knowledge" is simply one state among others that share the same imperfect footing of ultimate correctness or truth. The "science of the day" must be presumed inaccurate no matter what the calendar says. We unequivocally realize there is a strong prospect that we shall ultimately recognize many or most of our current scientific theories to be false and that what we proudly flaunt as "scientific knowledge" is a

> tissue of hypotheses—of tentatively adopted contentions many or most of which we will ultimately come to regard as quite untenable and in need of serious revision, or perhaps even abandonment.[9]

It must be noted, however, that although present science is not perfect, still it is better than that of our predecessors. So, let us take what we have, especially if experimental results are positive and if things get done. It seems to me that science should be judged on the ground of the *concrete* results it achieves, and not on a priori presuppositions. It is in my view a metaphysical—and unwarranted—assumption to claim that science will *never* be completed and perfect. This thesis is based, in the last analysis, on the assumption that after all we will never be able to reach the so-called reality-in-itself.[10]

It is fashionable, nowadays, to claim that past scientific theories are false, but this is not correct either: Newton's mechanics, for instance, is not false. Kuhn claims that a theory losing out to another one is displaced and rejected, but this is incorrect, too. In the physics departments, courses on classical mechanics, classical optics, and so on, are still taught. So it turns out that the theories in question are not displaced, but *partially rewritten*. The semantic views—like Van Fraassen's, for example—miss a very important point: physicists always use more than one global theory. Not only that: usually competing global theories are accepted, but scientists—unlike philosophers—see no problem in this situation, and keep their basic realistic attitude. If we look at the history of science, Rescher says, we find continuous overcomings and replacements. But we see only *some* failures, and not continuous ones (after all, overcomings and failures are quite different things). Rescher's point of view is partially justified, but does not offer sufficient ground for claiming that science will never attain a correct picture of reality: it does not follow *logically*, since it is just a bet we make on the basis of our past experience. After noting that ideal science is neither "something we have got in hand here and now" nor "something towards which we are moving along the asymptotic and approximative lines envisioned by Charles Sanders Peirce," he in fact concludes that "perfected science . . . is an idealization that exists 'outside time'—i.e., cannot attain actual existence at all. It lies outside history as a useful contrast-case that cannot be numbered among the achieved realities of this imperfect world."[11] As I hinted at previously, however, I maintain that it is a metaphysical assumption to claim that science will *never* be completed. Is this representation correct?

The argument has some good empirical evidence—related, as we just noted, to past experience—that supports it, but cannot said to be valid from the logical point of view. In order to reach the conclusion (c) *Science will never be completed*, we need in fact at least two very strong premises: (a) *There is something intrinsically unknowable in reality*, and (b) *Human beings, by definition, cannot know something which is intrinsically unknowable*. This statement, however, is different from the reasonable assertion that perfection does not belong in this world. In my opinion we cannot rule out—from the logical point of view, at least—the possibility that what Steven Weinberg calls "the final theory" will be achieved sooner or later, nor can we exclude that one among the many theories about the ultimate structure of matter (superstrings, loops in the sense of the Ashtekar-Smolin-Rovelli's theory,[12] etc.) currently on the scientific market will be proved to be correct. We can, I repeat, foster as many doubts as we wish about such possibilities, but holding firm that we are not in a position to give a strong metaphysical answer—either positive or negative—in this regard. It may also be noted, however, that the logical admission of a possible, future completion of science is not incompatible at all with Rescher's regulative idea of the pursuit of truth.

Rescher's account of the scientific enterprise avoids the shortcomings of both a super-realism which claims that our present science gives us *the* correct and ultimate picture of reality, and an anti-realism which equates science to the picture of reality endorsed by any primitive tribe. Ironically, Rescher, who renders homage to idealism (although interpreting it from an essentially pragmatic viewpoint), seems to take science more seriously than many other contemporary philosophers who depict themselves as convinced realists. He defines his own position—*approximationism*—in the following way:

> While the theoretical entities envisioned by natural science do not actually exist in the way current science claims them to be, science does (increasingly) have "the right general idea." Something roughly like those putative theoretical entities does exist—something which our scientific conception only enables us to

> "see" inaccurately and roughly. Our scientific conceptions aim at what exists in the world but only hit it imperfectly and "well off the mark." The fit between our scientific ideas and reality itself is loose and well short of accurate representation. But there indeed is some sort of rough consonance. . . . The realism it [the present book] espouses is one of intent rather than achievement—a realism that views science not as actually *describing* reality but as merely *estimating* its character.[13]

While I agree with this view, which represents both what practicing scientists are actually doing and what natural science is all about, my uneasiness in evaluating Rescher's strong commitment to science's absolute incapacity of reaching truth somehow remains.

3. TRUTH AND THE NATURAL ORDER

There are no such things as explanations *in* nature. Explanations are meaningful only *in relation to* us, human beings. We like to question nature and to receive answers from it, but explanations are meaningful only as long as there is a mankind thriving on Earth. We should never forget, however, that nature exists anyway, independently of both us and the opinions we may hold about it. Starting from these premises, it is possible to verify that Kant's system is undermined by some strong epistemological inconsistencies.

The philosopher of Königsberg claims in fact, on the one hand, that our judgments necessarily presuppose the presence of thought. On the other hand, however, he admits that there are objects whose existence is, in some way, independent of thought and of all kinds of human mental activity as well. It may thus be asked at this point: Is it consistent, from an epistemological point of view, to claim both that (a) there are *noumena* and that (b) all our knowledge pertains to *phenomena*? I think that the answer is bound to be straightforwardly negative. But there is an even more basic inconsistency, because we must question the nature of the relation between known objects on the one side and thought on the other. *If* all known objects are related to thought in an essential manner, and *if* furthermore our judgments necessarily presuppose the presence of thought, *then* it follows that we will never be able to have knowledge of objects which are purportedly independent of human thought. The real point at stake is, in sum, the following: According to Kant, empirical data, although they are not created by thought, are not independent of thought either. But how is it possible to claim that there is a "something" which, though depending on thought as long as its existence is concerned, nevertheless is not created by thought itself?

This is the heritage of all the post-Kantian philosophical systems. Analytic philosophers and in general linguistically oriented thinkers, along with idealists, Marxists, and the representatives of many other philosophical schools, must face these Kantian inconsistencies. Most of the problems of contemporary philosophy thus stem from Kant's dualism, and not only Hegel and Marx but also Wittgenstein and Frege (not to speak of contemporary thinkers) make sense only if we take Kant's thought into account. Consider, for instance, Popper's famous thesis of the three Worlds. I claim that, following a nuclear conflict in which all mankind got destroyed, only World 1 would survive, while Worlds 2 and 3 would definitively disappear. A possible objection might be the following: *Who* can say that, since there would be no human beings to testify it? It is clear that there is no logical demonstration for it but, nevertheless, we can reason using *analogy*. As we said before, astronomy and paleontology give us good insights in this respect, suggesting that the natural world is independent of mankind, while the historical, social, and psychological dimensions are not.

It should be clearly understandable by now that we are back—once again—to the problem of the relations between epistemology and ontology. Kant, in fact, tried to reconcile a (rather vague) realist ontology with a basically idealist epistemology, but any such attempt is bound to failure. The only way out in this case is given by the adoption of an idealist ontology that can match the idealist theory of knowledge, a step that Hegel, for instance, consistently made. The logical positivists and the linguistically oriented analytic philosophers, when facing this problem, usually choose the idealist path, although officially endorsing an empiricist stance. As is well known, Marx criticized Hegel for his reification of thought from the human subject. But I claim that we should criticize Wittgenstein, Frege, Quine, and many other contemporary authors as well for their reification of language from the human subject.[14]

While it may be justified to state that the social realm is mind-dependent, the natural realm is not. The existence of society implies the existence of mankind, while the existence of nature does not: think again of what we said above about the Popperian three Worlds. Philosophers often deem it possible to infer natural entities from conceptual ones and, given this fact, even the creation of nature by men becomes a logically obvious conclusion. But this is simply false. Man does not create the ultimate components of reality and, without the previous existence of nature, even his creative capacities—which are referred to the social realm—could not be explained. Thought, like anything else, is just a part of nature and possesses no existence outside it. The same may obviously be said of language, which has replaced thought in the contemporary analytic tradition.

Note that the position we have outlined thus far is not a form of materialism, because it does not take matter to be the ultimate component of reality. It is a much more modest stance, which limits itself to assert the existence of *something* which is essentially independent of human mind. This means that when the physicist Steven Weinberg takes fields to be the ultimate components of reality,[15] we do not charge him with being an idealist, because in no case does he claim that fields are mind-dependent. Clearly his statement must be empirically verified, and future scientific research will tell us if he is right or not. As long as philosophers are concerned, the problem of the ultimate structure of reality is a matter pertaining to scientific investigation, and Weinberg's stance is all right, from our philosophical point of view, because the existence of a mind-independent reality is clearly stated in it.

Modern and contemporary materialism has made a serious mistake in this regard, ignoring the history of science. No doubt mankind's beliefs about the structure of reality have changed a great deal with the passing of time: they quickly become outdated. What is never outdated, instead, is the belief in a mind-independent reality, and just *this* belief is the cornerstone of a well-founded scientific realism, because the fact that science has not yet been able to reveal the ultimate structure of reality does not affect it at all. It is clear, though, that philosophers as such have no particular a priori insight in this regard: this is the root of all Kantian-like mistakes. So I have nothing to object to reality being essentially made up of fields or superstrings. The essential point I want to make is rather that reality, no matter how it is structured, exists independently of human mind/thought/language, which, in turn, are nothing but particular manifestations of reality as such. What I claim is that should all thinking and speaking human beings suddenly disappear, reality would still be there. And we must remember once again that the theses of those who take nature to be mind-dependent are not consistent with scientific discoveries, because natural science shows us that nature existed when no human mind was present. As a matter of fact, we cannot even foresee whether there will be human beings in the future. What if an asteroid hits the Earth, a physical—and not logical—possibility which cannot be easily discarded? This certainly has nothing to do with human thought, language, and conceptual schemes. Men are not gods: they cannot eliminate the mind-independent dimension of reality and, despite what Hegel and many analytic philosophers claim, reality precedes logic and *not* vice versa.

Obviously, the theoretical dimension plays a creative role in our representation of reality, even though theories must eventually match the structure of the real world in order to be valid. By creating theories we reproduce reality, but the construction of theories is nothing but abstraction. Our theorizing and conceptualizing could not even occur were we not to start from a natural reality which is a necessary input. Furthermore, those who claim that there can be truth in a world devoid of men, even though no one would be able to recognize it, are in my opinion wrong. Truth is not a Platonic idea, but just a human device, while the order of things is not. Truth is correspondence between thought and reality: if one of the two is missing, there is no truth left. Since I deem it correct to assert that, if compared to the whole history of the universe, the presence of mankind on the Earth is not very important, we are bound to conclude that "truth" is simply a word, a series of letters put together. Truth is something we put forward in order to represent the way in which the natural world is ordered. The disappearance of mankind implies the disappearance of our representations of reality, but reality itself and its intrinsic order would not be affected by such an event.

4. THE KANTIAN DESCENT OF TODAY'S ANTI-REALISM

As we already observed several times in this work, a striking feature of the contemporary debate between realists and anti-realists, which is thriving in current metaphysics and the philosophy of science, is the Kantian flavor of many anti-realist arguments. The question to be asked now is: Are the philosophers involved in this debate aware of such a descent and, if so, to what extent? I argue that, although many authors recognize some kind of debt to Kant, the awareness of the Kantian heritage widespread in the current realism/anti-realism debate is not as strong as it should be. And this is particularly true within the analytic tradition, which, although originally inspired by strong empiricist standards, subsequently turned to a sort of linguistic idealism that is shared—often unconsciously—by most of its representatives.

It is not difficult to track these developments back to Kant's own words. If we consider again, for instance, the previously quoted pages of the *Critique of Pure Reason*[16] where he puts forward the example of the hypothetical inhabitants of the Moon who—according to his opinion—can be defined as "real" only when they are eventually known by us, we are right at the heart of Kantian thought. The philosopher of Königsberg, in fact, hardly denies that there is a reality independent of our knowledge of it (which, after all, might be considered as the ultimate *source* of knowledge itself). The real problem at stake is, rather, that what he calls "the world in itself" is unknowable. What he denies, in other words, is human beings' capacity to describe and know this purported independent reality. And it is well known that this situation is due in his view to the fact that for expressing our judgments we use a priori concepts and categories, which have nothing to do with an external reality as such. Kant's transcendental idealism thus says that our judgments are universal and necessary but, on the other hand, they can never reach reality in itself. Note, however, that this purported universality and necessity of our judgments is not based on empirical grounds, but is rather the outcome of a transcendental deduction that is highly problematic.

Having already remarked that these original Kantian theses are questionable today, given modern science's results, it is possible now to go a step further. Are we allowed, for instance, to draw a distinction between existence, which as such is absolute, and our knowledge of existence, which is not? In other words, we would like to know: (1) whether reality exists independently of our experiencing it or not, and (2) whether we are in a position to claim that there is a reality independent of our mind even when we get to know a preexisting reality only afterwards. According to Kant, this would clearly be impossible. He asserts in fact that objects, being never given in themselves, but only in experience, "have no existence outside it."[17] But we have to be extremely careful here, since Kant's philosophical genius cannot be easily trapped. Soon after, in fact, he readily admits that there may be inhabitants on the Moon "although no one has ever perceived them," so allowing his puzzled reader to imagine that, after all, he is willing to admit some kind of absolute existence. But—as we know—this is by no means the end of the story. Kant's concept of existence gets to the surface in a very clear manner in the next few sentences of the *Critique*, where he goes on to say that despite the fact that it must be admitted that there may be inhabitants on the Moon independently of any [human] perception, this "only means that in the possible advance of experience we may encounter them."

To put it briefly, we are told that something, in order to be real, must stand in connection with human perception or—even more clearly—with our "actual consciousness." This is the core of Kant's Copernican revolution and, indeed, here we find the roots of almost any subsequent development in the history of modern and contemporary epistemology and ontology. Man no longer is one out of many parts of reality, but becomes the center around which the whole of reality rotates. Kant soon after hastens to specify that (A) existence for us, that is existence which stands in connection with a perception (in accordance with the laws of empirical advance), is by no means the only kind of existence we can think of. (A) pertains in fact to the appearances which, as mere representations, are themselves real only in perception. But we also have the so-called (B) things in themselves, which are not supposed to stand in connection with perceptions in order to exist. And this is precisely the problem we have to face.

"To call an appearance a real thing"—Kant tells us—"prior to our perceiving it, either means that in the advance of experience we must meet with such a perception, or it means nothing at all." So, here two kinds of existence are assumed which, in the Kantian vision, cannot be identified. Let us call (A) "perceptual existence" (PE) and (B) "absolute existence" (AE). The problem is, however, that only PE is defined in clear terms, while AE is not and, as a matter of fact, we do not even know what to do with AE in the Kantian system: it is a sort of undefined concept perennially surrounded by a thick fog which prevents us from seeing what really lies behind. We know that the things in themselves are there, and that furthermore AE pertains to them. But there is nothing else to say except—perhaps—for the rather obscure notion that after all things in themselves are the "raw material" which is the *source* of our perceptions.[18]

I have argued elsewhere, both in this work and in a previous paper of mine,[19] that this Kantian dualism has been extensively adopted by the contemporary analytic tradition with only one caveat: for the analytically oriented authors "language" replaces "thought," although playing the same role (and it might be argued as well that the "language" of the analytic tradition is not *very* different from Kant's "thought"). We will return to this important point later on: now I would like to stress, instead, that the post-Kantian philosophy immediately realized the inner tension existing between PE and AE. Let us see why.

On the one hand, Kant clearly states that thought is essential to our knowledge in the following sense:

1. Whenever we want to pronounce a judgement on the world, we need not just thought alone, but something stronger: *interpretative* thought.

But, on the other hand, Kant's vision presents yet another strong presupposition:

2. Although (1) is certainly true, it is true as well that, at some ontological level which is not clearly specified, there are "objects" whose existence is totally *independent* of our thought, be it interpretative or not.

At this point one may note that the situation is getting really difficult to handle. Because it is clear that we have here a bifurcation at the ontological level between PE and AE while, at the epistemological one, we have an essential part of reality (that is, just that part which theoretically is the source of human perceptions) which is ultimately *unaccessible* to our knowledge. This Kantian outcome is no doubt dramatic, and explains why subsequent philosophers felt an irresistible need to solve it in a way or another. In other words, why should we assume that there are mysterious and unknowable entities called *noumena* while admitting, at the same time, that our knowledge has only to do with *phenomena*?

Let us try to summarize in a very sketchy manner what the implications of such a situation are. Given the account we have just outlined, it becomes impossible to determine the mutual relations—if any—between thought and the so-called *noumena*. So the argument goes:

a. as long as an object whatsoever is *known* to us, it gets known through thought;
b. but this in turn means that *no* object can be known to us independently of thought;
c. question: where do the known objects come *from*?

At this point we have two possible answers at our disposal: either (1) we eliminate *noumena*, claiming that thought is the only source of human knowledge; or (2) we save *noumena,* freeing them, however, of the rather mysterious characteristics they have in the Kantian system and claiming that *they* are the real source of human knowledge (but, then, it is obvious that they no longer are *noumena* in the Kantian sense, but just objects of common sense experience *external to*—and thus, to some extent, *independent of*—human thought).

It is important to note, anyhow, that there is no way of preserving this same system in the original form devised by its creator. Taking step 1 we follow the example of Hegel and of classical idealism. Taking instead step 2, we are bound to adopt some kind of materialism (like, for instance, the one endorsed by the Marxist tradition). We may well claim, thus, that most of modern and contemporary philosophy is nothing but a reexamination of Kantian epistemological and ontological dualism, and the outcomes of such a reexamination depend on which side of the Kantian dilemma we choose to foster.

Before closing this section we need to mention again a very important problem, with which we already dealt in the previous sections of this chapter. Kant's dualism—and consequently his famed "Copernican revolution"—can be understood only by taking into account his particular (and enormously successful up to the present day) conception of space and time as a priori faculties of the knowing subject. As is well known, for Kant space and time are not something objective and external to the human mind, but rather modalities by which the human mind itself organizes the—

alleged—surrounding world. Bertrand Russell expressed this state of affairs with his usual brilliant wit in the following way:

> Space and time [for Kant] are subjective, they are part of our apparatus of perception. But just because of this, we can be sure that whatever we experience will exhibit the characteristics dealt with by geometry and the science of time. *If you always wore blue spectacles, you could be sure of seeing everything blue. . . . Similarly, since you always wear spatial spectacles in your mind, you are sure of always seeing everything in space.* Thus geometry is a priori in the sense that it must be true of everything experienced, but we have no reason to suppose that anything analogous is true of things in themselves, which we do not experience.[20]

No doubt this genial Kantian intuition is fascinating and magnifies the human mind as the true *creator* of reality. Mankind's perennial eagerness to establish itself as the real master of the universe has been exalted beyond measure by Kant's philosophical system. If it were true that space and time spring from our internal faculties, men would indeed become the dominators of a reality which, given the above-mentioned assumptions, would be nothing but a mere projection of themselves. The problem, of course, lies in the conditional form of the statement: "*If* it were true." Did someone actually show that mind has an a priori structure? Do we have any *empirical* evidence supporting this thesis? In my view, we do not. The only thing we have is Kant's transcendental deduction of the categories in the *Critique of Pure Reason*. But that is not empirical evidence at all: it is just an outstanding piece of philosophical work which, however, can be questioned and rejected. Unlike other authors,[21] I think that the real problem at stake in this case is not to verify whether transcendental idealism is coherent but, rather, if it is *empirically founded*.

My claim is that philosophy cannot continue to deal with space and time as if these concepts belonged to philosophy alone, without a very large part of them belonging to the natural sciences. We must nowadays take into account, instead, what science tells us in this regard, and this is true especially if we work in the field of philosophy of science. Kant himself, after all, would certainly agree with such a stance, given the fact that he based his theory of space and time on Euclidean geometry on the one hand and Newtonian mechanics on the other, that is, on the two leading scientific theories of his age. Contemporary science—imperfect and subject to errors as it may be—discovered that space and time do not spring at all from our internal faculties, but are rather characteristics of physical reality itself (and, no doubt, this modifies the whole picture at our disposal a great deal). The growing uneasiness of practicing scientists towards the persistent and pervasive Kantian influence on the philosophers' approach to reality is thus easily comprehensible.[22] Of course, we all know that there are very significant exceptions, as is shown by these remarks due to Hans Reichenbach:

> Kant had no proof that we would never arrive at a totality of observations which could not be ordered in the frame of his a priori principles and which would make experience impossible, at least experience in the Kantian sense. . . . The postulate that experience in the frame of the a priori principles must always be possible is the unwarranted assumption of Kant's system, is the undemonstrable premise on which his system hinges. That he does not explicitly state his premise shows that the search for certainty made him overlook the limitations of his argument. I do not wish to be irreverent to the philosopher of the Enlightenment. We are able to raise this criticism because we have seen physics enter a stage in which the Kantian frame of knowledge does break down. . . . The new physics has shown that we can have knowledge outside the frame of the Kantian principles, that the human mind is not a rigid system of categories into which it packs all experiences, but that the principles of knowledge change with its content and can be adapted to a much more complicated world than that of the Newtonian mechanics.[23]

Nevertheless, the "Kantian flavor" which I have mentioned several times in this work is still dominating contemporary philosophy. And this is really puzzling, because it means that most philosophers of our day are just unable to historicize Kant, while they pretend to teach scientists the right method for historicizing the great figures of the history of science. To explain this fact, we must perhaps assume that Kant's system is viewed by most contemporary thinkers as a sort of "eternal philosophy," but what kind of justification may such an assumption really have? Furthermore, this situation is likely to worsen the previously mentioned gap between philosophers of science and practicing scientists, which has continued to grow deeper in the last few decades. In a recent book of his, Henryk Skolimowski writes:

> How is it possible that in spite of the notorious unreliability of our senses the physical laws, which are based on sense data, are final and irrevocable? [Kant] concluded that knowledge of physical laws is not arrived at via the mechanism of the senses, but via something that makes it imperative that this knowledge is both intersubjective and irrevocable. This "something" is the structure of the mind with its fixed categories, which the physical laws only reflect. . . . Kant had too astute a mind not to realize that, if we grant to physical science the claim of its universality over everything there is, then we dwindle to utter insignificance. . . . There is a transcendental element in Kant's conception of the human race. Although separated from religion, the human is regarded as, in a sense, sacred. . . . Physics explores only the surface, the phenomena; the "real things," "things in themselves," the *noumena*, are beyond the grasp of physics, beyond our understanding. . . . He was aware that a completely physical universe, which is completely knowable and describable in terms of physical laws, leaves us little comfort and little meaning, too. . . . We no longer hold such a conviction for we consider all knowledge to be revocable and tentative. Had Kant possessed our insight, or our hindsight, he would have no doubt thought up different solutions to both problems. His case is illuminating for two reasons. One is that great minds may conceive of marvellous solutions . . .and quite another is that even the greatest minds are at mercy of the assumptions of their times.[24]

But *why*—I wonder—should we think that if we grant to physical science the claim of its universality over everything there is, then "we dwindle to utter insignificance"? This stance, which has been endorsed by so many philosophers, seems to me both unwarranted and unjustified. Recognizing that man is just a part of nature does not necessarily lead us to such frightening conclusions; on the contrary, it is likely to offer a better understanding of man's position in the universe. As a matter of fact, naturalism can give us a picture of reality which avoids the inconsistencies hidden in transcendental idealism.

5. RESCHER'S PRAGMATIC IDEALISM

An original position within contemporary philosophy is the one adopted by Nicholas Rescher, who openly declares his idealistic stance. Throughout his works, Rescher maintains that a mild form of idealism, such as the one he endorses, represents the best way of dealing with philosophical issues. In this context I am particularly interested in analyzing his theses about the current realism/anti-realism debate, and I will therefore take mainly into account his book on scientific realism,[25] although referring, occasionally, to another work of his. First of all, however, I would like to find a definition of the term *idealism* broad enough to be generally acceptable, and the meaning of this move will become clear later on, when I shall try to show that Rescher's adoption of the term *idealism* may to a certain extent be detrimental to a correct understanding of his own positions.

The reason that prompts me to take this terminological detour is mainly historical. Idealism, in fact, has a story that is at least as long as the history of philosophy itself, but there is no doubt that we—living in the twentieth century—largely identify idealism with the classical formulation that Hegel gave of it, Schelling and Fichte being just a step behind. Furthermore, as is well known, idealism has been flourishing throughout continental Europe up to the present day,[26] and let me only remember that in Italy, for instance, the neo-idealistic systems of Benedetto Croce and Giovanni Gentile have dominated the philosophical scene up to the 1960s (and still count on many followers). The problem is that, as far as philosophy of science and more generally scientifically oriented philosophy are concerned, idealism has a "bad" story. Unfortunately labels, in philosophy, are often much more important than they should be. Philosophers of science tend to see the idealists as their natural "enemies," and the history of modern and contemporary philosophy offers good ground for claiming that they have at least some good reasons for thinking that way. To someone who grew up in Europe, Hegel's classical idealism (or even more recent forms of neo-idealism) has a very precise meaning. It means claiming (a) that reality depends on—or is created by—human thought, which in turn is a manifestation of a superior Spirit, and (b) that science is a secondary and less important kind of knowledge, because only pure philosophical speculation tells us how the world really is. Rescher's use of the umbrella-term *idealism*, thus, can be somehow misleading, because we will soon verify that there is no trace of antiscience in his works. But let us now take the brief terminological detour we mentioned before and, in particular, let us try to verify whether a reasonably general definition of idealism can be found.

If we take any philosophical dictionary it may be soon found out that idealism, no matter the many forms, trends, and subtrends into which it is divided, can be reduced to a very few basic propositions or axioms which are common in various degrees to all those who define themselves idealists. I extrapolate now from a well-known philosophical dictionary[27] a limited set of such propositions:

1. matter is somehow the manifestation of Spirit, however Spirit is defined;
2. matter can never be independent of mind and thought;
3. physical objects can have no existence apart from a mind which is conscious of them;
4. we could not possibly know unexperienced physical objects;
5. the existence of physical objects coincides with our perceiving them, because we could not conceive as existing in abstraction from our sense-experience any of the qualities we ascribe to them;
6. object implies subject, but subject also implies object, though subject is ultimately prior.

Obviously these six basic propositions may be articulated in different ways, but the core of idealism can be fairly reduced to what they state. Our task is now to verify to what extent Rescher's conceptual and pragmatic idealism fits into this very general picture, particularly taking into account the question of scientific realism. Note, first of all, that if we take idealism to be defined by proposition 1, Rescher is hardly an idealist. Idealism is in fact a typically monistic philosophy, in the sense that it chooses to dissolve the matter-element of the couple *matter/spirit* into the spirit-element, thus prompting evident doubts about the origins of matter itself. Starting from now, however, we will abandon the term *spirit*, replacing it with *thought*. It can be shown that Rescher is not a supporter of monism in the sense previously mentioned because, after all, he never claims that thought *creates* reality or that reality may be *completely* reduced to thought. His position is subtler, as the following quotation demonstrates:

> The idealist position maintained here holds that the things of nature are standardly envisaged in terms of a conceptual machinery that is substantially imbued with mind-involving elements. Such an idealism seemingly invites the tempting prospect of imagining an *an sich* reality, a natural world conceived of in ways totally denuded of mind-involving concepts, and so strictly objective and mind-independent. But this conception of reality *an sich* needs closer scrutiny. . . . The key factor here is the distinction between noumenal (= extra-phenomenal) reality, about which a great deal can be known in scientific theorizing, and *an sich* (= "purely objective," altogether mind-independent) reality, about which precious little can be known from the point of departure of the standard conceptual scheme we deploy in science and everyday life alike. . . . While the idea of a "mind-independent reality" is not wholly vacuous in import, it represents a *via negativa* that does not take us very far, and is in any case irrelevant for any information regarding "our reality" (because radically discontinuous with it). But just what is involved in such a reorienting of "reality" from the sphere of the altogether mind-independent towards that of an at least partly mind-conditioned truth? More must be said regarding the nature and the credentials of such a concept of reality not *an sich* but *for us*.[28]

The essential fact to be noted here is that Rescher is not denying in this context the existence (considered in a properly ontological sense) of a reality which, since is not created by us, is ipso facto external to us. What he wants to point out is, rather, that the categories we deploy in order to shape our experience are themselves put there by us; this is tantamount to claiming that the objects we get to know can be known by us only *in relation to* our conceptualization. We clearly have the Kantian descent still at work in this case, because even for Kant there is an empirical reality which, although not created by thought, cannot be considered totally independent of thought. And this in turn can lead us in two different directions: (1) it may mean that objects, whose existence can be ascertained only through thought, have no ontological status independent of thought itself; or (2) it may mean that we can get in touch with those same objects through thought simply because it is a fact pertaining to *nature* that we get to know the world this way. I find conclusion (2), but not conclusion (1), legitimate. But it is perfectly clear, in any event, that this process could not even occur if there were no raw material onto which our conceptualization is actually mapped. So, we have here a dualistic position—and not a monistic one. And this also gives us the first hint toward developing a thesis I will discuss later on: the fact that in my view Rescher's epistemology is mildly idealistic, while his ontology is not.

It might be difficult, however, to match an idealistic epistemology and a nonidealistic ontology. Kant, as we saw before, certainly tried to do something of that sort, and this explains his notion of "thing in itself," while classical idealism soon got rid of the problem, adopting a very clear idealistic stance even at the ontological level (and I insist that we can say much the same about most present-day, linguistically oriented philosophers). But the case with Rescher is certainly different, because in his book on scientific realism we find plenty of evidence that he has no intention of adopting a full-fledged idealist ontology, as appears evident in reading his own words:

> Seeing that a pragmatic line of approach pivots the issue on what is useful for us and productive for us in the context of our evaluatively legitimated aims and purposes, we return to the characteristic theme of idealism—the active role of the knower not only in the constituting but also in the constitution of what is known. To be sure, this sort of idealism is not substantive but methodological. *It is not a denial of real objects that exist independently of mind and as such are causally responsible for our objective experience.* Quite the reverse, it is designed to facilitate their acceptance. But it insists that the justificatory rationale for this acceptance lies in a framework of mind-supplied purpose. For *our* mind-independent reality arises not *from* experience, but *for* it.[29]

It is interesting to note that Rescher defines the picture he gives us as "an object-level realism that rests on a presuppositional idealism at the justificatory infralevel," which seems to justify my claim that he attempts to put together an idealist epistemology with a basically realist ontology (about which latter, though, as happens with Kant, we are not in a position to say much). But we can find many other passages by Rescher in which an ontological realism clearly transpires, such as, for example, the following:

> In the light of this analysis, the present discussion has a mixed or mediative character. It is a realism all right, since it acknowledges a realm of ontological mind-independent existence. But since it stands committed to a cognizability-in-principle standard of real/true/actual it is also a conceptual idealism which holds that existence is to be explicated in mind-referring terms of reference.[30]

At this point there is even more to say because Rescher, when he asks himself, "Is man the measure?" answers with a firm "No," claiming that we have no reason at all to think that our human reality is the only one there is, and this anti-Protagorean stance leads him to recognize that "Neither astronomically nor otherwise are we the center around which all things resolve" (and this position—I claim—is both anti-Kantian and anti-idealistic). Let us note, eventually, that he does not make the typically idealistic move of blurring any distinction between epistemology and ontology either, although recognizing that a boundary-line of separation between the two must be preserved only "insofar as possible."

Having offered what seems to me sufficient evidence for concluding that Rescher is not an idealist at least as far as *ontology* is concerned, I now turn briefly to his epistemology. He tells us that "realism is an *input* into our investigation of nature rather than an output thereof. It does not represent a discovered fact, but a methodological presupposition of our praxis of inquiry."[31] And he goes on, claiming that "insofar as realism stands on this pragmatic basis, it does not rest on considerations of independent substantiating evidence about how things actually stand in the world, but rather it is established by considering, as a matter of practical reasoning, how we do (and must) think about the world within the context of the projects to which we stand committed." Subsequently we are told that "it is clear that this sort of commitment to an objective realism harks back to the salient contention of classical idealism that values and purposes play a pivotal role in our understanding of the nature of things."

It seems to me that, if we carry out a careful scrutiny of Rescher's statements, his commitment to idealism is no stronger than his readiness to accept realistic conclusions about the relationships between thought and reality. Note, in fact, that it is certainly correct to claim that we do—and must—think about the world within the context of the projects to which we stand committed, but the truth of this statement is after all anchored to the recognition that our projects would *not* be the same if reality were different from how it actually *is*. What I would like to stress about Rescher's theses is, in sum, that the utilitarian role he envisions for the commitment to a mind-independent reality does not necessarily clear the path for idealism but, on the contrary, is perfectly compatible with a reasonable realism. And by this expression I mean a realism that, while holding firm that there is an external reality which by no means is created by the human mind, still overtly recognizes that we get in touch with that same reality directly through thought and conceptualization, and indirectly by means of induction and scientific instruments (as is the case, for instance, with paleontological discoveries about our Earth when mankind was not present on it). A realist should not be worried in accepting such a position, as long as his proponent drops the Kantian input about alleged a priori categories and faculties which would allow us to build up reality. But my claim is even stronger, since I maintain that not even a materialist should consider such a position unacceptable as long as, of course, he is a sophisticated materialist (i.e., one who does not identify matter with tables and chairs, but is ready to accept—to use a paradoxical expression—a rather "immaterial matter," such as the one envisioned by current subatomic physics).

It may well be—as Rescher claims—that initially realism is an input into our investigation of nature rather than an output thereof. But what if we just reverse the order of the process, saying that our sensory apparatus meets since the beginning of our life *something else*, which can be identified as external reality, and *only* then do we proceed towards theorizing and conceptualizing? We could as well reconcile Rescher's position with the one just put forward, claiming that our image of reality depends on the human mind (from the epistemological viewpoint), while reality does not depend on the human mind at the ontological level. Man's conceptual apparatus is no doubt *intentioned to* the world. But while I can perfectly understand a world with no human mind which intentions it, I believe it is quite hard to imagine a human mind with no world to be intentioned. This seems to offer us enough ground for claiming that reality is *ontologically* prior to our intentioning it. And even if it were true that we can know the world only through categories put forward by our mind, this fact would not change this ontological priority at all: it has a considerable bearing only on epistemology. If we thus reexamine the six basic axioms given before, I deem it correct to say that none of them is entirely consistent with Rescher's theses. Certainly (1), (2), and (3) would not be endorsed by him, while (4), (5) and (6) probably would, but only *sub condicione*, that is, by putting forward many conditions which inevitably weaken their range. So, it seems to me that Rescher's conceptual and pragmatic idealism is much closer to American pragmatism than to European idealism, and thereby shows that pragmatism does not need to be interpreted idealistically. In the following chapter, we shall have more to say on the relationships between ontology and epistemology in Rescher's thought.

CHAPTER 4

Ontology and Apriorism

1. ONTOLOGICAL REALISM

As we hinted at in the last section of the preceding chapter, Rescher's work can be deemed as a serious endeavor at both reviving the idealist perspective within contemporary analytic philosophy and making it compatible with a scientifically oriented outlook on reality. I argued before, however, that Rescher's conceptual and pragmatic idealism cannot be defined as a "true" idealism because, while adopting an idealist stance at the epistemological level, he nevertheless endorses a clearly realist position at the ontological one, and this means that a nondogmatic realism can actually be compatible with his position. Furthermore, his particular brand of idealism gives rise to a rather flexible system, which can face criticisms without giving up its premises. Now I would like to make clear and possibly deepen this remark by taking into account, besides the two volumes quoted in the third chapter,[1] a more recent article of his, also devoted to the theme of conceptual idealism.[2] I believe in fact that the issues raised by Rescher in these works are important for both a correct understanding of the current realism/anti-realism debate in metaphysics and the philosophy of science, and for the clarification as well of the manifest (although often undeclared) "idealistic turn" which took place within analytic philosophy in the last few decades.[3]

Just to understand the real problem at stake here, let us single out a statement that may be found at the beginning of Rescher's paper on conceptual idealism:

> Idealism, broadly speaking, is the doctrine that reality is somehow mind-correlative or mind-coordinated. However, the specifically *conceptual* idealism which concerns us here stands in contrast to an *ontological* doctrine to the effect that mind somehow constitutes or produces the world's material."[4]

As I see it, precisely this ontological doctrine is the core of both classical idealism and the many forms of neo-idealism which have been thriving up to the present day within Western philosophy. A short historical inquiry, in fact, will make us recognize that idealism is the philosophical doctrine according to which the existence of the world (reality) is either logically or causally dependent on the existence of some kind of mind.[5] This mind, in turn, may be human or divine, but such a further distinction is not important for our present purposes. Rescher thus clearly rejects from the beginning the most characterizing feature of both classical idealism and contemporary neo-idealism, claiming explicitly that his conceptual idealism stands in contrast to the aforementioned ontological doctrine. The importance of this move needs not to be stressed, because it is well known that the basic antiscientific attitude endorsed by the classical idealists and the contemporary neo-idealist thinkers stems precisely from it: why should we deem natural science important if it deals with a second-level reality, that is, with a reality that is "secondary," being merely a creation of the human (or divine) mind (or spirit)? According to this view, only philosophy—or some type of philosophical theology—can actually reveal to us the real "essence" of the world, given the fact that the mind/spirit creates reality as a counterpart to itself.

In Rescher's conceptions, however, we find nothing of this sort. Not only is physical reality irreducible to mind; he claims, furthermore, that natural science gives us a fairly adequate picture of reality, although warning against the diffused temptation of conceiving our actual science as a perfect and final representation of reality itself.[6] No doubt, neither Hegel and the founding fathers of classical idealism, nor Bradley, Croce, and other more recent neo-idealist philosophers would find themselves at ease with such statements. And the picture gets even clearer if we add that, as mentioned above, the ontological independence of the physical world is explicitly admitted by Rescher. The problem, then, is to find a rationale for what he often calls "intimations of idealism," and this will be our task in the following section.

2. EPISTEMOLOGICAL IDEALISM

Rescher's epistemological idealism is stated as follows:

> [Conceptual idealism] maintains that an adequate descriptive characterization of physical ("material") reality will always involve an implicit reference to mental operations—that some commerce with mental characteristics and operations always occurs in the explanatory exposition of the real world."[7]

After making clear that a "real world" exists, one that is neither identified with the mind nor created by our conceptual apparatus, Rescher goes on to adopt a position which the realist can accept *sub condicione*, that is, posing conditions we shall take into account later on. His main aim, in fact, is to distinguish between

A. Nature-as-we-understand-it, and
B. Nature itself.

Is this distinction ontological or epistemological? To answer this question, we must trace a line of separation between ontology and epistemology and, since Rescher is inclined to accept such a division (unlike many other thinkers nowadays), we are entitled to claim that ontology aims at discovering what kind of entities make up reality ("what there is," to use a Quinean expression), while epistemology's task is to discover the principles by which we get to *know* reality. The importance of this step is in my opinion crucial.

Going back to the aforementioned distinction between (A) and (B), in fact, it is clear that we are after all allowed to make an ontological distinction between them, but with an important caveat: only (A) is mind-dependent, in the sense that a hypothetical complete disappearance of mankind would make it disappear too, while (B) is neither created by the human mind nor dependent on it. Men can understand nature itself (i.e., B) in a great number of different ways, and the history of astronomy and cosmology gives us a clear picture of this situation. Our distant ancestors believed that the Earth was at the center of the universe, while a few centuries ago it was discovered that it is simply a part of a vaster system of planets rotating around the Sun. More recently the solar system turned out to be a very tiny part of an even vaster system called a galaxy, and our galaxy eventually was discovered to be a rather small part of an extremely complex system made up of billions of galaxies. And the story is by no means concluded, since the correctness of the current standard theory about the origin of the universe—the Big Bang model—has not been demonstrated with absolute certainty.[8]

So nature-as-we-understand-it really changes. Our vision of it is quite different from the vision held by the ancient Greeks, the men living in the Middle Ages, the founding fathers of the United States of America, or even the men living in nineteenth-century imperial Britain. The question to ask at this point is the following: What follows from this from an *ontological* point of view? In my opinion the answer is not as difficult as many contemporary philosophers seem to believe. No matter how much our visions of nature change, nature itself does not change for *that* reason. Some thinkers nowadays claim that history of science shows the impossibility of adopting an ontology which is absolute and mind-independent, and even say, for instance, that the world of the ancient Greeks is just different from our world. This may certainly be correct, provided we give a clear definition of the term *world* (reality). If they mean that nature-as-the-Greeks-conceived-it and nature-as-we-conceive-it are just different *pictures* of the world that have an ontological import on their own, one cannot but say that they are right. If instead they mean that the ancient Greeks' beliefs about their gods, the structure of Earth and heaven, and so forth had a direct bearing on nature itself, then this is patently false. The ancient Greeks happened to live on an Earth which is also our Earth. The only difference between their Earth and ours—both conceived of in ontological terms—is given by time conceived as a characteristic of physical reality which, together with space, puts heavy constraints on what we may or may not do. Today we can, for instance, travel from Europe to the United States in a reasonably short time, filling up a portion of space which was deemed almost insuperable by our ancestors. We cannot, instead, travel back in time to the Rome of Julius Caesar, nor can we fill up the portion of space-time which separates our planet and galaxy from other distant galaxies of the universe. These possibilities (and impossibilities) in turn are just a contingent fact, the consequence of how the universe—and the human beings who are part of it—are made.

Resuming the previous distinction between (A) and (B), I think we are now in a better position to judge the status of both its constituting elements. (A)—i.e., nature-as-we-understand-it—is subject to continuous changes due to our ever deepening comprehension of (B)—i.e., nature itself. Science is precisely the attempt at getting a good knowledge of nature itself, and it does not matter very much if we believe—with Rescher—that science will never be completed, or if we believe instead—with the physicist Steven Weinberg[9]—that contemporary science is very close to establishing what he calls the "final theory." It is not important for our purposes because, in any event, we are not jeopardizing the distinction between an ontologically mind-independent reality and the representation we build up of it, given the particular and contingent features of our conceptual apparatus.

In order to enforce what we just said, I would eventually like to mention two additional examples of Rescher's basic ontological realism. He states, in fact:

> In attempting answers to our questions about how things stand in the world science offers (or at any rate, both *endeavors* and *purports* to offer) information about the world. The theory of sub-atomic matter is unquestionably a "mere theory," but it could not help us to explain those all too real atomic explosions if it is not a theory about real substances. If I hypothesize a robber to account for the missing jewelry, it is not a hypothetical robber that I envision but a perfectly real one. Similarly, if I theorize an alpha particle to account for that photographic track, it is a perfectly real physical item I hypothesize and not a hypothetical one. Only real objects can produce real effects. . . . The theoretical entities of science are introduced not for their own interest but for a utilitarian mission, to furnish the materials of causal explanation for the real comportment of real things. . . . What science says is descriptively committal in making claims regarding the real world, but the tone of voice in which it proffers these claims always is (or should be) provisional and tentative. . . . Thus our inability to claim that natural science as we understand it depicts reality correctly must not be taken to mean that science is a merely practical device—a mere instrument for prediction and control that has no bearing on describing "the nature of things." What science says is descriptively committal in making claims regarding "the real world," but the tone of voice in which it proffers these claims always is (or should be) provisional and tentative.[10]

The picture is even more complete if we recall that, in Rescher's view: "Instrumentalism puts the cart before the horse. As far as the working scientist is concerned, scientific theories do not exist for the sake of prediction and control, but the other way round—

prediction and control are of interest because they serve to monitor the adequacy of our theorizing about objective reality."[11] It seems very clear to me that here we have the explicit admission that there is an external reality which is not created by our mind or thought, however strong the epistemological constraints that our mind or thought imposes on that reality may be. We have, in sum, neither the classical idealism of the past century, nor the neo-idealism of our century, nor eventually a form of linguistic idealism such as the one I criticized in the preceding chapters of this work.

3. ARE THERE A PRIORI FACULTIES?

At this point there is nothing that prevents us from agreeing with Rescher when he argues that "an adequate descripitive characterization of physical ('material') reality always involves an implicit reference to mental operations." This statement, in fact, is tantamount to recognizing the well-known fact that we get in touch with reality—including ourselves as a part of it—through conceptualization, which, in turn, can be expressed by means of thought plus language. And, having granted this fact, we must now try to verify what kind of consequences may be drawn from it. Since Rescher clearly states that mind is responsible for nature-as-we-understand-it, that is, (A), and not for nature itself, that is, (B), we have eliminated from the outset one of the major critical points of both classic idealism and contemporary neo-idealism. Idealism comes back, instead, when he claims that mind has a *constitutive* role in nature; that is, indeed, a characterizing point of every form of idealism, including the one endorsed by those analytic philosophers who assert that language has a constitutive role in the world. Soon after, however, Rescher clarifies at best his position, adding that this constitutive role of the mind must be taken neither in ontological nor in causal terms, but rather from a hermeneutic viewpoint.

So we face a situation of the following kind. On the one hand, mind does not produce nature, but on the other, we always need the analogy of mind for conceiving nature itself: our reference to the external world is always mind-correlative. Here we have another crucial passage which involves the presence—or not—of a priori factors in our knowledge of reality. A short reflection, in fact, makes us understand that, starting from the aforementioned considerations, two different paths can be taken according to the definition of "conceptualization" one is ready to endorse. In other words we can choose a Kantian-flavored solution, saying that our conceptualization of the world takes place because in our mind a *creative element* is present, that is, some faculties or categories that cannot be found in nature and are an exclusive feature of the human mind. These faculties or categories shape external reality in a certain way that is common to all human beings *qua* human beings: for instance, they allow us to conceive of things ordered in causal relations, placed in space and in time, and so on. And right here idealism acquires a great role, because if this Kantian-flavored position is correct, then it is obvious that a person is not a mere part of reality but is entitled to claim a much more important status. He or she becomes somehow a *creator* of reality.

What really strikes one in this position is that it is quite hard to understand *how* this creative human capacity fits in the order of nature and why it came into being at all. We had on our Earth a natural process of evolution which, although not completely known thus far, can nevertheless be described with sufficient precision. In this process, humans come on the stage relatively late—according to cosmic standards, of course—and gradually (i.e., not at once) acquire a predominant position thanks to their capacities, which are in turn a product of natural evolution. These capacities, however, must not be seen as a *violation* of the previous natural order if we want to be consistent with a scientific picture of reality; they are, rather, a higher step in the development of this same natural order, and conceptualization is thus a tool that allows humans to get both a better picture and a reasonable control of what lies outside them. Note that if we take this second path, the human capacity of conceptualizing becomes something natural and explainable, although our comprehension of its historic origin and concrete evolution is by no means complete. But, on the other hand, there is no cogent reason for claiming that we will never be able to understand when and how humankind acquired the above mentioned capacity; I think it it reasonable to view it as a constantly growing addition of new skills to previous ones.

There is, obviously, a big *Why?* behind all this. Even scientifically oriented thinkers like Karl R. Popper and Konrad Lorenz[12] recognize that it is quite difficult to answer questions of this sort:

Why did evolution take the direction we know? Why did life begin at all on the Earth? and so on. At this point we are bound to stress a very important fact indeed. If mind is allowed to be not only a complex natural structure that reflects an external reality, but also an ontological agent which possesses a priori faculties of the Kantian sort, one can reasonably ask: *Where* do these a priori faculties come *from*? By endorsing an aprioristic and Kant-style view, we are somehow forced to conclude that a significant violation of the natural order took place at an indefinite moment of time in the past. *Something* happened, even though we do not know precisely what, that endowed a finite being with superior capacities and, thus, we need to admit the presence of an agent who caused the aforementioned "something." Granting that we can make as many hypotheses as we like, some kind of teleological principle viewing evolution as the means by which consciousness arises would now be justified and, to point out just one example out of many, the conceptions of the French thinker Pierre Teilhard de Chardin would thus be vindicated (at least in principle, if not in detail).[13] According to this vision, only the presence of a priori mind-equipped beings would in fact justify the presence of a physical universe which, in their absence, makes no sense at all. To use one of Rescher's examples, it might in fact be asked: "Why should there be any smelling of roses if there is no mind-equipped being who can smell them?" Or, to put it in a more straightforward way, who cares about a universe in which there are roses but no human subject who can smell them? But this reasoning is in my opinion unjustified, because it is tantamount to saying that the universe is significant *only* as long as humans exist. I would like to make it clear that, in following this path, we are abandoning the domain of philosophy to enter a new field of inquiry, and, at this point, at least three choices are available:

1. we may accept the violation of the natural order as an assumption based on some type of religious faith;
2. we may accept the violation of the natural order as an assumption based on faith, although maintaining a difficult equilibrium between faith and science;
3. we may refuse any violation of the natural order, sticking to what science says.

In any event, it is clear that if we accept the aforesaid violation of the natural order without having sound empirical evidence in its favor we are just abandoning the domain of rational inquiry, and this does not seem to be a justified philosophical stance. Furthermore, we can note that, in order to accept position 1, we do not even need to endorse a traditionally religious position. 1 is in fact compatible with the presence of a sort of Hegelian Spirit and with the philosophical theology that is today so popular and successful after having been revived by many authors belonging to the hermeneutic tradition.

This is not Rescher's position, however, although he endorses the presence of a priori factors in our knowledge. We are then called to verify in what sense his conceptual idealism is different from other idealist stances that can be found in the history of Western philosophy. He depicts his outlook as claiming that the key properties we ascribe to physical things are *relational*, in the sense that our conceptual scheme works in such a way that the properties we ascribe to those things always involve some type of reference to mental operations. According to my interpretation, these words mean that, no matter how physical reality—whose independent existence is beyond any doubt—is truly made, we always need a conceptual scheme to get in touch with it; note that the conceptual scheme is needed not only to ascribe properties to stars and galaxies, or tables and chairs, but even to the components of our own body (hands, feet, etc.). This body of ours is in fact part of the external reality, while the properties we ascribe to its components would, according to this picture, be assigned to them by our mind or conceptual apparatus.

Just take the following example: Does a particular tree have the shape it has because our mind attributes a shape to it, or is this shape something that our mind somehow reproduces, picking it up from nature itself, that is, (B)? Consider a particular (not a possible, but just a real) tree, say x. It happens sometimes that lightning burns a tree, changing its shape, and suppose furthermore that lightning—say y—actually burned x, changing its shape. In what sense is this change of shape—due to a natural phenomenon like the lightning, y—dependent on men's minds? I do not see any plausible idealistic answer except the Kantian one, which, as I have already pointed out, does not agree with what science tells us.

What I want to stress is, in sum, the following fact: The representatives of classical idealism, and Hegel in particular, recognized that idealism must—just to be consistent with itself—eliminate any separation between ontology and epistemology. Only by claiming that physical reality is a product of some kind of mind/spirit will it avoid the tension between a realist ontology and an idealist epistemology, and this is exactly what Hegel did by getting rid of the Kantian *noumena*, while on the other hand Marx took the opposite step in order to obtain a consistent materialist system. Now, it is interesting to see how Rescher solves the old problem of the contemporary presence on the stage of a realist ontology and of a materialist epistemology. Are they truly irreconcilable even when taking his particular form of idealism into account?

4. AN ALLEGED INCOMPATIBILITY AND ITS SOLUTION

At this point, Rescher distinguishes between "conceptual mind-involvingness" and "explicit mind-invokingness" and picks up two different items in order to illustrate this distinction: books and dreams. A dream is explicitly mind-invoking because dreams can only exist where there are mind-equipped beings to do the dreaming, while books, being physical objects, seem at first sight entirely non-mental. Nevertheless, books are mind-involving, because there could be no books in a world where minds had never been in existence. And he goes on to claim that:

> Even if there were no mind-endowed beings, there could certainly be naturally evolved book-like objects, objects *physically indistinguishable from books as we know them*. . . . A book is, by definition, an artifact of a certain purposive (i.e., communicative) sort. . . . To be a book is to have *writing* in it, and not just *marks*. . . . To explain adequately what a book is we must thus make reference to writing and, in turn, ultimately to minds.[14]

We may summarize Rescher's distinction by saying that there are immaterial objects like dreams and worries—let us call them O_1—which are clearly mind-invoking, and material objects (human artifacts) like books—let us call them O_2—that are clearly mind-involving. In no way could they exist without the presence of human beings. It is in my opinion clear, however, that a third kind of object is needed to have a complete picture of the situation and to justifiy the separation between ontology and epistemology that Rescher ultimately accepts. This third kind is formed by physical objects—we may call them O_3—which are neither human artifacts nor mind-dependent. If we take a typical O_2 object like a book, in fact, it is easy to see that it is made of paper (another O_2 human artifact), but in order to produce paper we need a physical O_3 object like cellulose, which can hardly be deemed to be mind-dependent (it gets *in touch with* our mind, and this is a completely different story). So, while it is correct to claim—as Rescher does—that "a world in which there neither are nor ever have been minds can contain objects physically indistinguishable from our books and nails, but books and nails they could not be, since only artifacts created for a certain sort of intelligence-invoking purpose can correctly be so characterized," I think that we must add something else. Leaving aside "possible" entities like book-shape objects indistinguishable from our books, what we need is a realm of physical objects that are not made up by us, that is, a mind-independent external reality.

Here, however, even a further step is mandatory. Let us think of reality-as-we-conceive-it—the aforementioned (A) in Rescher's previous distinction. We have already established that it cannot be identified with (B)—nature itself—which, as such, is mind-independent. What are we supposed to do with (A)? A naive realist would most likely answer that we do not need it: being a product of our conceptual apparatus, we can just rule out the possibility that it has any ontological value whatsoever. I do not find this position acceptable, though, and for a very simple reason: How can we deny the existence of entities like dreams and worries, which are clearly *there*, although not in the same sense as physical objects are? The fact of the matter is that, as Aristotle claims, "being" can be talked of in many ways. We live in an extremely complex reality. Besides facing an external world which is not produced by us, we also build up a reality which is the product of our mind. On the one side (i) we cannot identify those two kinds of reality, but on the other (ii) we can neither deny existence to one of the two, because it is a fact that they both exist, nor (iii) can we divide them with too rigid a boundary. There is in fact *continuity* between (A) and (B): neither can a dream be separated from the subject who dreams it, nor can a worry be detached by the subject who holds it.[15]

The best thing to do, then, is to attribute to them a different—but not totally separated—kind of existence. It is by no means correct to claim that my worry that the book I am writing will not be published is "unreal"; it is as real as the desk I am writing on because both of them are present, so that both are endowed with some kind of existence. In order to clarify the issue, let us draw the following distinction:

1. There are O_1 objects like dreams. Their existence, however, is strictly dependent on human minds, in the sense that, should mankind disappear, there would no longer be dreams.
2. There are O_2 objects like books. Their existence is still tied to human minds as far as we consider both their contents (ideas and theories contained in the books, etc.) and their shapes (their book-like appearance), but not when we take into account their ultimate material ingredients which belong to an external and mind-independent reality.
3. There are O_3 objects like trees from which the cellulose we use to make paper (and thus books) ultimately comes. Their existence is not dependent on human minds.

Quine, of course, would disagree with such an "overcrowded" ontology but, as previously noted,[16] I think that Ockham's razor, which he so often applies, poses too many problems to be adopted uncritically. Quine is still prisoner of the positivist dogma according to which "real" is only what can be empirically observed, a dogma that contemporary science has shown to be untenable. If reality is complex, ontology is bound to be complex as well: we cannot oversimplify reality according to our personal opinions. So we have two kinds of existence, a conceptual existence, E_1, which is predicated of conceptual reality: it is tied to our conceptual apparatus and is bound to vanish along with it. Secondly we have an absolute existence, E_2, which is predicated of external reality and is not bound to vanish along with our conceptual apparatus. It should be clear, however, that both kinds of existence are *real*, the only difference being that E_1 is tied to worlds in which human beings with their minds are present, while E_2 has a broader range, since it is relative to *any* kind of world, either inhabited or not by human-looking and mind-equipped entities. Note, however, that I did not employ the notion of "possibility" because, in this context, it would be a source of dangerous misunderstandings. In mentioning worlds devoided of human presence, I do not mean "possible" worlds in a metaphysical (Leibnizian) or logical (modal) sense, but just worlds ontologically as real as our Earth: as we remarked before, both paleontology and astronomy give us examples of such worlds, and this is a sufficient reason for rejecting the ontology of classical idealism and of contemporary neo-idealism (which is different, as we already saw, from Rescher's).

At this point we no longer have the incompatibility between a realist ontology and an idealist epistemology. Granting the mind-independent existence of reality itself, we can admit without problems not only that (1) mind shapes reality-as-we-conceive-it, but also that (2) only through mind can we get in touch with reality itself.

Unlike the other entities that form reality itself, in fact, we are mind-equipped, which means that we get in touch with external reality through conceptualization. However, since this faculty needs not to be assimilated to the a priori faculties envisioned by Kant and then taken up by the classical idealists, but may instead be explicable as the outcome of a natural process of evolution, we do not need to assume, either, that reality itself exists as long as it gets in touch with our perceptual apparatus. We are indeed creators of some type of reality (i.e., reality-as-we-conceive-it), and in this sense dreams, worries, literature, philosophy, and so forth are all products of the human mind. But certainly we do not create trees, stones, atoms, planets, etc. On the contrary, we can even formulate the following hypothesis: While it is reasonable to assume that the structure of the world-as-we-conceive-it reflects the structure of the mind, I see plenty of reasons for claiming as well that the structure of our mind reflects that of a world mind it must *comply with* in order to survive along the path of a long evolutionary process. Mind thus simply becomes, according to this vision, the terminal (by now) point of an evolutionary (natural) process which is still under way. Our world (i.e., nature-as-we-conceive-it) would not even exist without mind's capacity of conceptualizing and, as a matter of fact, we cannot even imagine a different way for getting in touch with reality itself. Certainly no "vision of the world" would then be possible.

But the point I want to stress is that the physical universe did not wait for our vision (or better, visions) of it in order to exist. The evidence of physical worlds devoid of human presence is so overwhelming that I do not really see how this argument might be rejected. After all, it is simply a contingent fact, and not a necessary one, that human life evolved on our planet and, at the present stage of scientific reasearch, we do not even know whether something comparable to human life and mental activity developed on some planets belonging to our or other galaxies. Reality, as it may be, is divided into parts. I am just a part of reality, and the tree is another part. Our final constitution is essentially the same, being reducible to ultimate components. Now, it happens that within me some chemical reactions take place, while within the tree other chemical reactions take place: life, in sum, manifests itself in different ways. This line of thought allows us to naturalize all the alleged a priori faculties that an important part of Western philosophy has put forward in order to explain our knowledge of the world.

5. A COMMON OBJECTION AND SOME NEW INSIGHTS

One of the most common objections to the picture I have just given is: "Even admitting that what you say is true, *what* can we possibly know of a purported reality itself that lies outside the range of our conceptualization? Maybe it is out there, but it is not relevant at all for our purposes, since we can say nothing about it." I have argued elsewhere[17] that this is simply a matter of *attitude* towards reality. Just proceeding by analogy, think how difficult everyday life would become should we start considering "real" only what we are able to perceive here and now. Fortunately, our basic and inherited attitude towards the world that surrounds us is very different. I *know* that there is a corridor outside my office even though, the door being closed, I cannot actually see it. Similarly I *know* that, although my body is now located in the American continent, there are nonetheless other continents in which billions of human beings are actually living. This is not only a logical inference, but something more. It is a basic attitude that stems from conceiving ourselves as parts of a larger reality, one that ultimately expands up to the boundaries—if any—of the physical universe. And it is also worth mentioning that this same basic attitude is the cornerstone of scientific research: no comprehension of the world would be possible starting from the assumption that conceptualization is a sort of prison we cannot escape from. Neither an astronomer looking for distant galaxies nor a physicist looking for subatomic particles would, at that point, be justified in their endeavours.

Rescher somehow endorses the common objection we have just mentioned but, before taking this fact into account, I would like to consider another important characterization of conceptual idealism according to his own words:

> Conceptual idealism sees mind not as *causal source* of the materials of nature, but as furnishing some of the *interpretative mechanisms* in terms of which we understand them. It maintains that we come to cognitive grips with nature on our own terms—that is, in terms of concepts whose makeup involves some reference to minds and their operations. Conceptual idealism is predicated on the view that reality as we standardly conceive—in terms of material objects characterized by empirical properties and relations with one another in a setting of space and time—is actually in some respect involved in making reference to the operations of the mind. As *this* sort of idealism has it, the mind understands nature in a manner that in some ways reflects its own operations in some fundamental respects.[18]

These statements are crucial because what we have here is on the one hand the explicit ruling out of any *causal* production of nature by the mind, and on the other a strong underlining of the autonomy of mind as such, which is thus supposed to give a central and creative contribution to the construction of the world. Rescher is also stressing that he is not talking of any particular mind, be it mine or that of the friend I met on my way to the office this morning. *Which* world are we talking about, however? The dependence of nature-as-we-conceive-it on mind is necessary in the sense that it is systematic in nature; as we already saw, it depends on the way we—*qua* human beings—are made, which obviously means that there is nothing "voluntaristic" in this situation: we just cannot change our way of conceiving the world that surrounds us. But in what sense is nature itself dependent on mind?

At this point we must face a double possibility. Let us start again from the point that Rescher takes for granted: there are both a mind-independent nature itself and a mind-dependent nature-as-we-conceive-it. Mind plays a pivotal role in producing nature-as-we-conceive-it which, in turn, is the only way we have at our disposal for getting in touch with nature itself. As I see it, the double possibility is the following one:

a. Either we interpret this role of the mind in a *naturalistic* way, saying that it is a product of natural evolution and its structure largely reflects that structure of an external reality it must match in order to guarantee the survival not only of the single human individuals, but of the whole species; or

b. We choose a Kantian path granting the presence of a priori categories such as causality, space, time, etc., that allow us to be creators of reality in a strong sense: we do not only create nature-as-we-conceive-it, but also have a somewhat creative role in the setting of reality itself.

My claim is that, starting from Rescher's basic premises, neither (a) nor (b) can be ruled out: it is a matter of how one chooses to pursue his basic idea of the distinction between nature itself and nature-as-we-conceive-it. What is clear, however, is that conceptual idealism becomes a mild form of idealism perfectly compatible with non-naive realism if we choose (a), turning instead into a Kantian-flavored and stronger form of idealism if we choose (b). In some parts of his paper,[19] Rescher seems to endorse a Kantian position of this type:

> [Kantian idealism] holds that, even though there is a world whose nature and existence is neither logically nor causally dependent upon the existence of any human mind, the properties of that world are knowable not as they exist apart from experience but only as they are experienced. Although rational beliefs and scientific knowledge may be epistemically virtuous, no methods of inquiry have the capacity to reveal the world as it exists in itself.[20]

This is particularly evident during his discussion of space and time, when he claims that "all *positioning* of things (be it in relating them to one another observationally or with reference to a coordinate-placement scheme of some sort) involves a mental process. The very idea of location and placement in particular identifiable positions is conceptually mind-invoking."[21] But, as I see it, these statements do not necessarily involve the adoption of a typically idealistic stance: everything depends on what we conceive space, time, and our own human minds to be. The fact that at this moment my wife is six hours distant from me in time and about five thousand kilometers distant in space does not depend at all on my conceptual apparatus placing her that way. It is due, instead, to the structure of a physical reality of which space-time is one of the key components, a structure that contains both me, the desk I am writing on, my spatio-temporally distant wife, and the spatio-temporally even more distant Andromeda galaxy (and, last but not least, my mind). In other words, I see things located in space and time because reality is just *like that*, and not because my mind places them that way. What about saying that we see things the way we do because we are inserted in an Einsteinian space-time physical structure? Would we still conceive the *same* reality if space-time had just a different structure?

It might of course be objected that I always use a definite reference point, an "I" in terms of which everything else is seen, judged, thought, spoken of, and so on. This is undeniable but, again, the consequences we are entitled to draw from this fact depend in turn on what this "I" is taken to be. If the "I" is not some type of transcendental subject , but just a concrete individual who uses language and thought as historically evolving natural tools, then the picture changes a great deal: for sure, we are no longer compelled to conceive of our mind as a sort of prison, since this—to some extent unpleasant—role is left out to a physical reality of which I am, along with my fellow human beings, just one out of many components. Picking out an example made by Rescher himself[22] and already used previously in this chapter, the smelling of roses, aside from the romantic meaning which is attributed to it by human beings, is basically a fact of chemical reactions. In a mind-denuded world both human beings smelling the fragrance of roses and the complex set of meanings attached to it by human beings would no longer be there, but the chemical reaction, yes, it would *still* be there, as we know from the fact that chemical reactions took place on our Earth when humans were not yet present on it (and many of the consequences of those far-distant chemical reactions still affect us). Another example may be the following: the fact that night and day succeed each other seems at first sight mind-involving, but it is not. Mind-involving is our division of time in a twenty-four-hour-long day. If someone asks, "What if there is no human being thinking about night and day?" we can answer that what we call "night" using our language is something which exists even if there is no one attributing a name to it. It is not, in other words, a question pertaining to our conceptual schemes or to linguistic conventions, but a feature of the universe, and those authors who adopt a Kantian stance seriously risk running into the reification of the human mind.

The point I would like to make clear, thus, is one that Rescher himself raises, but that risks in my view being overshadowed if someone takes his idealist epistemology more seriously than his realist ontology. He in fact readily admits that:

> The conceptual idealist thus has no vested interest in denying that mind and its functioning may be somehow *causally emergent from the process of matter.* His point is simply that our standard rational account of the world—its material sector specifically included—is given in terms of reference that are at bottom mind-involving. . . . This position . . . is quite *compatible with the idea that mental functioning has its material basis and causal origins in the realm of physical process.*[23]

These remarks—as we have already noted—are important indeed, because they trace a clear line of demarcation between Rescher on the one hand and traditional idealism on the other. Once we hypothetically admit that mental functioning has both its material basis and causal sources in nature itself, many misunderstandings that Rescher's position may induce in the realist (or even strictly materialist) philosophers are immediately cleared. At this point nothing prevents us from claiming that mind is a natural tool that reflects the structure of nature itself, a product of evolution meant to guarantee the survival of the most complex being evolution has ever produced on our planet. And nothing prevents us either from eventually abandoning that "reification" of thought and language that, taking them to be a priori faculties, has brought contemporary analytic philosophy into a blind alley.[24]

There remains only one point of disagreement between Rescher and a realist, and it is, needless to say, a Kantian point. Even this disagreement, however, is not as irreconcilable as it would seem at first sight, which confirms the basic flexibility of his system of conceptual idealism. Rescher makes clear at the end of his paper[25] that "things in themselves" as Kant envisioned them are untenable. The task of discussing them as a mysterious sort of objects is—he states—hopeless, and there is no separate ontological category of *noumena* that needs a special consideration. There is, I would like to add, no mysterious rose-in-itself that lies behind the rose-as-we-perceive-it and that—most likely—is the causal source of the Kantian phenomenon knowable to our mind. Rescher's claim at this point is that if we abstain from using the mind-invoking concepts put forward by our conceptual apparatus, among which he lists causality, spatio-temporal features, experientially accessible properties, and so forth, we can say very little about what is left after these mind-invoking concepts have all been laid on the carpet. "With all of our standard descriptive categories in abeyance," he says, "we are pretty well rendered mute with respect to the characterization of this realm, once we have presumed its existence and its existential essentiality to our familiar world."

My claim is that nothing prevents a sophisticated realist from taking such a picture for granted but, at the same time, the solution he gives to the problem is bound to be different from the one Rescher endorses. We have seen previously that there is only one ontology. Between nature-as-we-conceive-it and nature itself there is a genuine continuity, in the sense that our mind, being, along with thought and language, a natural product of evolution, cannot be something essentially different from nature as such. Mind's representation of reality cannot be based on a problematic a priori structure because, otherwise, we cannot even explain how come we *do* represent reality: representation would become at that point a mysterious process whose origins we will never be able to ascertain. Most of modern and contemporary philosophy has reached the questionable conclusion that there is an unbridgeable gap between the picture we get from our senses and the one that science gives us because, after all, there is an undeniable difference between the two. But this is simply a consequence of the fact that evolution has equipped us with a "picturing" structure that is useful for our survival as a species. In order to survive, we do not need to perceive the subatomic particles: the perception of common physical objects is more than sufficient for that purpose. It is, in sum, a mere fact of contingency if today we cannot observe many scientific entities, a contingency which rests, in turn, on both the characters of the aforementioned entities and those of the perceptual mechanisms we are endowed with.

Subseqently science was able to penetrate both the microscopic and the macroscopic structure of reality, and this is a success, not a failure, of our senses, which in fact have hitherto been "extended" to a degree unknown to our distant and less distant ancestors. Entities which were not detectable in the past centuries became perfectly detectable in our days, mainly thanks to the artificial devices used in current scientific practice. The difference between detectable and undetectable entities, thus, pertains to the epistemological level, and not to the ontological one. This view may be rejected only if the distinction between ontology and epistemology is rejected, too, and we already said that there is no plausible reason for endorsing such a stance (despite its present popularity).

There is no mystery lying beyond: why should we assume that what we do not yet know is *unknowable*? This is exactly the sense of what practicing scientists call the "search for the final theory," a search that will probably go on for a long time and which stems from human beings' natural desire to know a reality they are part of. There is no mystery in the chemical reactions that explain the fragrance of roses, and the fact that we do not yet know the ultimate structure of reality should not lead us to believe that there is a mystery there either (unless one wants to resort to the unwarranted Humean notion of the so-called hidden springs in nature). To assume—as Rescher does—that we standardly think of reality in mentalesque terms is correct and undeniable. By why is this so? Because this is the way we are made: it is a contingent fact that has nothing to do with a priori faculties. Rescher's solution to the problem is certainly different from the one we have just sketched, based as it is on the presence of a priori faculties in the human mind. But, still, I deem it reasonable to argue that the conclusions we have reached in this chapter are not essentially incompatible with the premises his conceptual and pragmatic idealism ultimately rests on.

It is worth noting, at this point, that Rescher, in discussing with me his conceptual idealism, added some new insights which, although they are not contained in his published works, seem quite important.[26] Their importance stems from the fact that they clarify at best his position and eliminate many possible misunderstandings. These insights make us understand that Rescher's ontology differs from the one we have sketched in the following sense. While we contemplate an ontology which includes O_3 objects like trees, he claims instead that, however real trees may be, the *conception* of "trees" that defines them as such is based on human ideas, which in turn depend on how we botanize nature in our conceptual scheme.

Rescher does not deny that in our experience we have two sorts of "there is a tree" responses: (1) those which are delusional (i.e., misimpressions, optical illusions, etc.), and (2) those which are veridical, which occur when we are *really* confronted by a tree (because we touch it, we cut it down, and so on). He points out, however, that the sentence *There actually are trees in the real world* is a somewhat oversimplified formulation for something which would—more correctly—run as follows:

> There occur (and, thus, *exist*) processes in the real world—processes which themselves nowise depend on the existence of minds, and of which we humans can have a rough and imperfect knowledge through natural science. These processes, in turn, nondelusionally engender the "here is a tree" response in duly prepared minds.

It follows that the trees we experience (and think of, and talk about) are something which, in the final analysis, do not exist *as such* (i.e., as conceived of by us), but rather as something like Locke's secondary properties. They would be, in short, the results of certain trends of interactions between mind-independent natural processes on the one side and human minds on the other. These processes go along their own (realistic) way, and causally engender the veridical beliefs that we entertain. We can thus admit that there are *real* trees which grow *real* leaves and produce *real* fruit. But—Rescher claims—this is still something of a shorthand. What really goes on are certain *processes* (partly but only imperfectly understood by us) which produce in us the (veridical) conviction that there are "trees."

The outcome is that Rescher's ontology, insofar as it is realistic, is an ontology of (imperfectly grasped rather than *an sich*) processes in the world, which in their imperfection and interaction with human minds produce an intellectual manifold of thing-characterizations. The fact is, however, that this causal production is always *mediated* by the brain-physiological basis for a conceptual scheme. All this means that the processes that lead us to the conviction that "there is an apple tree" operate independently of us, but the "appleizing" of these processes—i.e., the apple-envisioning response to them—always calls for the intervention of the mind. The causal basis of our apple-encounters is real and mind-independent, while the transmutation of that causal basis into what is recognizable as an "apple" always involves a mind. Rescher's ontology, according to his own words, is thus structured in the following way:

1. Mind independent processes
2. Minds
3. Causal products in those minds produced by the impacts of these mind-independent processes

Rescher neither rejects things or substances nor denies their reality, but sees them as ontologically subsidiary and derivative in a way that involves items 2 and 3, which renders them mind-dependent. We have, in sum, (a) a realistic ontology of physical processes that are somewhat amorphic because only imperfectly understood, and (b) an idealist ontology of substances or things.

No doubt these explications make the picture clearer and help us understand Rescher's position. Note, however, that they do not falsify our basic remark, according to which he joins a basically realistic ontology and a mildly idealistic epistemology. We may ask ourselves: Is Rescher implying that an extraterrestrial with a conceptual apparatus different from ours would see what we see as an "apple" in a different manner? Maybe so. But this has nothing to do with the processes which belong to the ontological structure of reality, and not to any kind of conceptual apparatus (be it human or extraterrestrial). The fact of the matter is that I, my conceptual apparatus conceived of in brain-physiological term, and the apple are made of exactly the same ultimate components. Clearly Rescher can maintain that there is a difference between (a) the conceptual apparatus and (b) its brain-physiological basis. So we come to the point that the "scheme" of the conceptual apparatus itself is something intrinsically different from its "content." But to what extent is such a clear distinction tenable? It may be noted that we do not want to reduce mind to "matter" (or vice versa); we are rather saying that no distinction of this sort is well grounded because there is only *one* reality.

Let us then resume the example of our Earth without human beings. No one would botanize trees; no one would call the trees "trees"; and no one would think about trees. We can even formulate a further hypothesis: should amoeba become the most complex living being on the Earth, instead of humans, there is a chance that it would perceive what we call "trees" in a different way (but this, of course, must be shown). What Rescher calls "processes," however, would still be there and, as a matter of fact, we can figure out an *ontology of processes* in which no human being is present. It follows, in my opinion, that the distinction between a necessarily realistic ontology and an epistemology which, in turn, needs not be idealistic may be maintained. Our conceptual scheme, or conceptual apparatus, or reference framework (in a Wittgensteinian sense), cannot be conceived of without—or somehow detached from—nature: it *is* a part of nature itself. We must *naturalize* our conceptual scheme if we want to make sense of it at all. If we think instead of it as something outside or transcending nature, then how do we explain its existence (presence)? It seems to me, in other words, that Rescher keeps a gap between thought/language on the one side and nature/reality on the other. But ontology, as we said before, is one: language and thought are just two components of nature out of many.

At this point, we are also bound to ask: What does the expression "duly prepared minds" mean? The only available answer, in my view, is that they must be prepared to "something," and the something we perceive *cannot* be intrinsically different from the something that exists. This is likely to be the secret of conceptualization: our concepts must somehow follow lines that are present in the processes of nature. *Why* should we introduce an unbridgeable gap between things/processes and concepts/representations? I am perfectly willing to accept Rescher's statement that we know very little about those processes, even though they do not lie behind (whatever "behind" means) nature: they are just *there*. Maybe our actual conceptual schemes are unable to reach a certain level of reality but, still, the history of science is one of ever deeper insights into both the micro- and the macro-cosmos. Furthermore, for reasons that were previously explained, I take for granted that Rescher's processes are not the Kantian *noumena*. Those are intrinsically unknowable and always lie behind *phenomena*, while there is nothing intrinsically unknowable in the processes as Rescher has them. We cannot rule out the possibility that in the future science will make us know what they really are, and there is thus no reason to believe that our conceptual scheme is *irreducible* to natural processes. If we make such an assumption, we are also bound to conclude that man is not a part of nature, and that our presence in the world, being inexplicable, is a violation of the natural order. There would no longer be continuity between men and nature, but just an *unexplainable* gap.

CHAPTER 5

The Road to Naturalism

1. RORTY AND THE REIFICATION OF LANGUAGE

In a recent paper of his, Richard Rorty says that what Gustav Bergmann called the "linguistic turn" was "a rather desperate attempt to keep philosophy an armchair discipline. The idea was to mark off a space for a priori knowledge into which neither sociology nor history nor art nor natural science could intrude."[1] This, of course, would explain well enough why linguistic analysis has become with the passing of time a sort of first philosophy, aimed at replacing precisely that metaphysics which the founding fathers of logical positivism gave up for dead. Rorty then goes on to claim that what Ian Hacking has defined as "the death of meaning" (i.e., the end of any attempt to make language a transcendental topic) cleared the way towards a more naturalistic way of conceiving language itself. Essentially, he says, we moved from Frege and the early Wittgenstein, who are the philosophers primarily responsible for holding the idea that language can be defined as "a clearly shared structure," to the second Wittgenstein (later followed by Quine, Davidson, and Rorty himself)[2] who instead gave up this idea.

To clarify the issue, we can thus claim that two opposite conceptions of language are at stake here. The first says that (i) language is something *self-explanatory* which, in turn, explains everything else. This means postulating type A objects (i.e., *unexplained explainers* like Platonic forms, Kantian categories, Russellian logical objects), in terms of which type B objects (the *explananda*) can be accounted for. The second conception, instead, claims that (ii) there is no actual distinction between type A and type B objects. All objects are on a par, but in a very special sense. Rorty resorts in fact to the Quinean-flavored similitude of the net and its nodes, about which more will be said later on. If we want to avoid the self-referential problem constantly faced by those who postulate a distinction between type A and type B objects,[3] then it is necessary to change the whole picture. Since according to Rorty "the Tractarian distinction between the available and effable world and the unavailable and ineffable 'substance of the world'" is wrong (and—I would like to add—the Kantian distinction between the available and effable phenomena and the unavailable and ineffable noumena has serious shortcomings, too), all we have to do is to give up the distinction altogether.

This is a nice strategy indeed. Since there is no longer ineffability and unavailability, all problems seem to be solved. But is this move justified? We have good motives to be suspicious here because, after all, Immanuel Kant must have adopted that kind of model for some reason, and Ludwig Wittgenstein himself struggled, up to the end of his life,[4] with the old problem of the gap between reality itself and our representations of it. Rorty goes on to claim:

> If we are allowed to say that type A entities are their own *rationes cognoscendi*, or their own conditions of linguistic accessibility—that they make themselves available without being related to one another or anything else—then we are faced with the question of why type B entities cannot themselves have this obviously desirable feature.[5]

Assuming—as Rorty does—that Wittgenstein's *Tractatus Logico-Philosophicus* can be read as a heroic attempt to save philosophy from naturalism by claiming that type A objects must be ineffable, that they can be shown but not said, and that furthermore they can never become available in the way that type B objects are, we observe that the price is too high to be paid. Why, after all, should philosophers think that such a *heroic* attempt is the core of their job? This is a question I keep asking myself when reading the earlier Wittgenstein and all those philosophers who desperately try to attribute a priori elements to human mind (and language) in order to prevent the conclusion that we do not belong to a separate ontological category in comparison to the rest of nature. It is much more reasonable to see human beings, along with their thought and language, as a mere part of reality that deserves no special type of consideration within the order of nature itself.

2. THE HOLISTIC SOLUTION

Let us check, then, the Davidson-style solution that Rorty proposes. At this point, in fact, he proclaims his faith in holism: in contrast to the assumption that there can be entities which are what they are, totally independent of all relationships between them, a Davidsonian (and also Quinean) holism claims that "all entities are merely nodes in a net of relations," which gives us a picture of the following kind:

> No intrinsically simple objects, no pictures, and no language. For if analysis could not end with such objects, then whether a sentence has sense would depend, *horribile dictu*, upon whether another sentence were true—the sentence which specifies that two simpler objects making up a composite stand in the relevant compositional relationship. But when one asks what would be so horrible about *that*, [the first] Wittgenstein has no obvious answer.[6]

So one may wonder: *What* is so horrible about that, after all? In my view the "horribleness" of such a situation is due to the fact that, despite all the differences that Rorty finds between the first and the second Wittgenstein, and between himself and the orthodox analytic philosophy, even in his vision our language becomes not only the arbiter of truth, but also the sole component of reality. As Ian Hacking has it:

> Rorty's version of pragmatism is yet another language-based philosophy, which regards all of our life as a matter of conversation. Dewey rightly despised the spectator theory of knowledge. What might he have thought of science as conversation? In my opinion, the right track in Dewey is the attempt to destroy the conception of knowledge and reality as a matter of thought and of representation. He should have turned the minds of philosophers to experimental science, but instead his new followers praise talk.[7]

In adopting such a picture, we are back, once again, to the old idealistic view that human beings cannot step out of their thought, the only difference being that "thought" must in our days be replaced by "language." Obviously, this is not the interpretation that Rorty gives us of his own stance. In his opinion, he—together with other contemporary thinkers like the aforementioned Quine and Davidson—chose to overcome the linguistic turn in order to adopt a sort of "conditions free" view of reality. "Availability"—he claims—"requires being related by something other than the relata themselves. We have opened up the question of why we ever thought that there was a problem about availability in the first place. We have thereby questioned the need for philosophy, insofar as philosophy is thought of as the study of conditions of availability." But can availability be so *easily* discarded?

Rorty adds that we must turn to "naturalism," the view that anything might have been otherwise, that there can be no "conditionless conditions" (we shall see later on that this view may in effect have some truth in it, provided we do not understand it in a Rortian way), and he ends up by claiming that historicism is just a case of naturalism so conceived.[8] It is worth noting, however, that his definition of "naturalism" is different from ours. We, in fact, take naturalism to be the thesis that any explanation, in order to be meaningful, must be brought back to a natural order in which all humans are inserted since they are born, and this natural (i.e., spatio-temporal) location does not depend on their personal will. Fetzer and Almeder claim instead that naturalism can be reduced to naturalized epistemology, and even attribute to Nicholas Rescher the view that the most privileged and reliable methods for acquiring an understanding of the world are those used by science.[9]

The solution that Rorty proposes, thus, relies on the standard interpretation of the second Wittgenstein, who according to his opinion dropped the idea of finding nonempirical conditions for the possibility of linguistic description. But what does the second Wittgenstein mean by *empirical*? This is by no means clear, despite the opinion of many interpreters of the Austrian philosopher's writings. According to Rorty, "He (the second Wittgenstein) became reconciled to the idea that whether a sentence had sense did indeed depend upon whether another sentence was true—a sentence about the social practices of the people who used the marks and noises which were the components of the sentence." This may be interesting from a philosophical viewpoint, but we are bound to ask ourselves whether the aforesaid conclusion does really reflect what the later Wittgenstein had in mind.

The fact is that Wittgenstein's texts do not lend themselves to a clear and straightforward interpretation such as the one endorsed by Richard Rorty in his article. I would like to single out, in this respect, a beautiful passage drawn from *On Certainty*, where Wittgenstein addresses the nature of our reference framework dealing with Moore's solution to the skeptical problem:

> It might be imagined that some propositions, of the form of empirical propositions, were hardened and functioned as chan-

> nels for such empirical propositions as were not hardened but fluid; and that this relation altered with time, in that fluid propositions hardened, and hard ones became fluid. The mythology may change back into a state of flux, the river-bed of my thoughts may shift. But I distinguish between the movement of the waters on the river-bed and the shift of the bed itself; though there is not a sharp division of the one from the other. . . . And the bank of that river consists partly of hard rock, subject to no alteration or only to an imperceptible one, partly of sand, which now in one place now in another gets washed away, or deposited. . . . The truths which Moore says he knows, are such as, roughly speaking, all of us know, if he knows them.[10]

Here we are at the very core of the later Wittgenstein's thought, a core which he reached just shortly before his death. What kind of sense can we make of Wittgenstein's remarks? An orthodox Wittgensteinian would most likely say that, in this context, our philosopher is dealing with the problem of the relations among (i) propositions, (ii) beliefs, and (iii) the framework within which we give grounds for our beliefs.[11] But let us explore instead an alternative path, without worrying too much about the faithfulness of our interpretation to the orthodox gloss of Wittgenstein's texts.

In *On Certainty*, as quoted above, we read that: "the river-bed of my thoughts may shift. But I distinguish between the movements of the waters on the river-bed and the shift of the bed itself; though there is not a sharp division of the one from the other." All right, we may answer, but imagine that thought, language, and our brain are one and the same thing, just expressed differently. Do we still need the beautiful metaphor of the river-bed? Our hypothesis is that there is no difference between us, the river-bed and the water, while Wittgenstein still talks of the "frame of reference" as if it were something clearly distinguishable from ourselves, stars, tables, and so forth.

The great issue at stake here is once again, in my view, the relation between ontology on the one side, and epistemology on the other. True, most interpreters would point out that such a distinction is untenable in Wittgenstein's thought, and in particular if we take into account the second phase of his philosophical parabola. But, notwithstanding this common opinion, I think we are here confronted with a great problem, namely, that of determining *what really is* the reference framework about which Wittgenstein so often talks, and which, furthermore, is supposed to be shared by all human beings *qua* human beings. He frequently says in his works that skepticism raises doubts when no questions can be asked, while Monk correctly describes his endeavor in *On Certainty* as one aimed at showing "The point at which doubt becomes senseless."[12]

But how can language—now taken to be a set of social practices—guarantee us against skepticism? Wittgenstein at this point resorts to the celebrated notion of the "language game," saying that there is no language game in which questions concerning the reality of the world can be asked. Wittgenstein is right in this respect, but *why* so? My claim is that we need to go a step further, gearing any linguistic game to something else (which, however, is not ontologically separated from the language game itself). If we take the expression *when no questions can be asked*, are we challenging (i) our scheme of reference, (ii) a reality as it is and could not be otherwise, or (iii) both, because they are exactly the same thing?

Let us go back to Wittgenstein's similitude of the river-bed. At the beginning he makes a move which is purely linguistic, claiming that (*a*) some (empirical) propositions hardened and began to function as "channels" for (*b*) other (empirical as well) propositions that remain fluid. The purpose of both *a*- and *b*-type propositions is to furnish a world-picture that, however, might be part of a "mythology." Suppose that the hardened *a*-type propositions are more or less the river-bed, while the fluid *b*-type propositions are the waters which are supposed to flow within the river-bed. The border line is not so clearly defined, just as it happens with the river-bed and the waters, but in any event we can always trace some kind of distinction. The *a*-type propositions are thus similar to the bank of the river that consists of hard rock, while the *b*-type propositions are like the waters that flow. Our referential framework consists essentially of *a* type propositions, whose validity is self-explanatory and such that they are not touched by any skeptical doubt: to question them, in fact, is tantamount to question our reference framework. Those *a*-type propositions form what we may define as the *fundamental* language game, on which everything else rests. And *who*—we might ask—established this fundamental language game? Wittgenstein's answer would most likely be that language itself established it. But, this time, language is no longer an abstract and Platonic structure shared by all members of mankind. It is, rather, a set of social practices that has no ultimate meaning beyond itself.

Rorty relies on this late Wittgensteinian picture to build his own picture, which, as we said, is in turn much indebted to Quine's and Davidson's conceptions. However, it is clear that the Wittgensteinian picture does not work very well if we take it at its face value. A whole set of questions comes in fact to our mind: (1) Does this picture imply that even nature is a set of social practices, and that, furthermore, it is only socially constructed? (2) Do we build the laws of nature socially? (3) Do we construct the ultimate components of reality science is looking for socially? And (4) is the second Wittgenstein arbitrarily applying the notion of "social practice" to the whole of nature? Such a move seems to me totally unwarranted (although being no doubt fascinating from a purely philosophical point of view, as Rorty and other thinkers have shown). As a matter of fact, the framework about which Ray Monk talks so extensively can ultimately be neither linguistic nor social. At this point we can see that, most likely, Rorty is wrong in dismissing the following intuition of the early Wittgenstein as it is contained in the *Tractatus*:

> If the world had no substance, then whether a proposition had sense would depend on whether another proposition was true. In that case we could not sketch out any picture of the world.[13]

Maybe Wittgenstein, especially in his first masterpiece, used expressions like "substance of the world" in a rather deceptive way, but I take the basic preoccupation expressed in the aforementioned quotation as a completely justified one.

What is, then, Wittgenstein's "frame of reference"? In my opinion it can be nothing but nature conceived of as the whole of reality, which, in turn, is not mind-dependent. On page 563 of his valuable essay, Ray Monk tells us that the opposite of the sentence "physical objects exist" is not false, but incomprehensible. Yet—one may wonder—are we talking about propositions here? The answer is bound to be negative, even though Wittgenstein did not dare to make the final and most important step: that is, abolishing any distinction between language and nature. It is certainly true that practice shows sense, but the real question is: *Why*? Because—it should be answered—we comply to nature. It is not a matter of being merely "tuned" with it, since we are just a *part* of the natural order.

In any event, we should also ask: What does it mean to say that Moore's statements have the peculiarity that their denial is not just false, but incomprehensible? As is well known, Wittgenstein thought that if Moore were to pronounce the opposite of those propositions which he declares certain, we should regard him as demented. Even in this case, though, we are bound to wonder: *What* does this mean? Ray Monk says in this regard:

> If the contrary of a proposition makes sense, then that proposition can be regarded as an empirical hypothesis, its truth or falsity being dependent on the way things stand in the world. But if the contrary of a proposition does not make sense, then the proposition is not descriptive of the world but of our conceptual framework; it is then a part of logic.[14]

In Wittgenstein's picture, then, we have (1) the way things stand in the world, and (2) our conceptual framework. There are, in turn, (1a) propositions that are empirical hypotheses because they describe *how* things stand in the world: their contraries make sense; and (2a) propositions that do not describe how things stand in the world: their contraries do not make sense. The first, (1a), describe the world, while the second, (2a), describe our conceptual framework. Therefore, (2a) are part of what the Austrian-born philosopher calls "logic." But I claim that at this point we still have a kind of dualism. If the aforementioned view is correct, there exists a clear separation between reality and our conceptual framework. It should be admitted, on the contrary, that there is no boundary line between the world and our conceptual framework, and that our language poses neither limits nor bounds, being itself a part of reality. "Physical objects do not exist" is just as comprehensible as "Physical objects exist." Contemporary physics allows us to conceive of a situation in which an extremely concentrated quantity of matter exists, with no physical object in the current sense of the term being present, and there is no logical contradiction in this picture. The river-bed certainly modifies the waters: it makes them dirty; and the waters, in turn, modify the river-bed: they change its course. But the dirtying of the waters and the shaping of the river-bed obey the same *natural* laws.

If the standard interpretation of the later Wittgenstein were correct, how could we possibly explain our capacity to think and talk? It is obvious that these faculties should be attributed to extranatural factors. But the fact of the matter is that we can explain sufficiently well how, why, and when human language was born. It *is* a social practice, but social practice itself—like language and thought—is not self-explanatory. The only way out is to conceive of reality as everything that *exists*, where this "exists" is self-explanatory and is in need of no further explanation. The danger in the standard interpretation of Wittgenstein's position is, in sum, to be led to believe that things are the way they are due to some linguistic game put forward by social practice. But such a position is unsustainable, since any linguistic game must be geared to reality just to make sense. The problem stems from the fact of seeing representation and reality as totally separate factors: exactly *there* the gap shows up. True, we always have the impression that our representations match reality only to a certain extent and not completely. But the difficulties arise because we tend to see reality as something external to ourselves, while men and their conceptual schemes are *internal* to reality, being just a part of it.

3. THE WEAKNESS OF RORTY'S VIEW

Subsequently, Rorty quotes Thomas Nagel as claiming that Wittgenstein's later thought depends on a position so radical that it undermines the weaker transcendent pretensions of even the least philosophical of thoughts. This position entails that any thoughts we can form of a mind-independent reality must remain within the boundaries set by our human form of life. Let us turn our attention, then, to operative scientists. Can we admit that the subatomic particles which physicists take into account belong to a "human form of life"? From a certain point of view it is justified to claim that the standards of physics are set up by the scientific community, which in turn is nothing but a subpart of the human community at large. The fact is, however, that this form of life puts us in a condition to glimpse the very way reality is made. I do not see how physical reality may also be seen as part of a human form of life; neither can the continuous discoveries in the macrocosmos lead us to believe that the universe is something of that sort. Wittgenstein's insight is deeply Kantian in nature: it does not deny that there is something external to our mind, but rather claims that this "something" may be grasped only through conceptualization. But if instead we take thought and language to be parts of nature just like anything else, including ourselves, it soon turns out that apriorism has no empirical base, and this is not due to the limits of our thought or language, but to the fact that nature is made in a certain way.

In my opinion Rorty is right in claiming that the criteria for correct or incorrect reasoning are not provided by some external realm of Platonic truths, while he is instead wrong in saying that those criteria are exclusively provided by ourselves, by "a *convention*, or a *use*, and perhaps our practical requirements." If we follow consequently this path, in fact, we shall be forced, sooner or later, to admit that reality is our *creation*. Maybe there is out there a reality which does not depend on our minds, but we can say nothing meaningful about it. And so it is perfectly justified to claim that Rorty (and Wittgenstein, if we endorse the standard interpretation of his position), are both adopting a typical Kantian stance. Their problems are those with which the philosopher of Königsberg felt bound to cope, and they can be solved only by admitting that our thought and language are not *that* important. We must take them as part of something bigger: how can one be scientifically oriented about everything but thought and language?

It is easy to verify, furthermore, that Richard Rorty's view has another serious shortcoming. Take his picture of reality conceived of as a net formed by nodes whose esistence is only given by their being in relation to each other. It is clear that (*a*) the nodes have sense *only* within the net, but it is by no means less clear that (*b*) the net has sense *only* in-so-far as it is formed by single nodes. The absence of even *one* of these nodes would make the net just *different* from what it is.

Rorty (along with Davidson) insists on the notion of "relation," as if it could explain everything. Davidson's account of human linguistic behaviour"—Rorty claims—"takes for granted, as the later Wittgenstein also did, that there are no linguistic entities which are intrinsically relationless. But—I claim—the ontological structure of reality is not given *only* by the relationships between entities. Besides being related to one another, the nodes of the net must *be*: their natural existence is at least as important as their relatedness and, as a matter of fact, it might be argued that something must first exist in order to be related to something else. In other words, we could never make up a net from *nothing*: the net is bound to be made of some stuff. And it is worth noticing that we have indeed neither empirical nor philosophical reasons to deny that those nodes are precisely the ultimate constituents of reality which contemporary physicists are looking for. I think that Rorty's account, although intriguing from the philosophical point of view, is too partial: it dissolves reality within a concept of "social practice" that is too broad and loose to explain anything. Rorty might obviously answer that we do not need explanations because there is nothing to be explained, but this vision is hardly tenable. The second Wittgenstein, whom he considers—along with Dewey—as his mentor, never claimed that there is nothing to be explained: maybe he said that we cannot explain everything we wish to, and this is a completely different story.

As Rorty himself recognizes, his account of reality is much indebted not only to the second Wittgenstein but also to Donald Davidson's refusal to distinguish between conceptual schemes and their contents. I believe that Davidson's refusal is—to a certain extent—justified, although my reasons for claiming this are certainly different from the ones he puts forward. Let us quote again Rorty's words:

> Whereas in the *Philosophical Investigations* Wittgenstein still toys with the idea of a distinction between the empirical and the grammatical, between nonphilosophical and philosophical inquiry, Davidson generalizes and extends Quine's refusal to countenance either a distinction between necessary and contingent truth or a distinction between philosophy and science. . . . Instead Davidson asks us to think of human beings trading marks and noises to accomplish purposes. We are to see this linguistic behavior as continuous with nonlinguistic behavior, and to see both sorts of behavior as making sense just insofar as we can describe them as attempts to fulfil given desires in the light of given beliefs.[15]

The situation can be sketchily summarized in the following manner. If (1) we *qua* human beings are just a part of nature, then (2) our conceptual schemes are a part of nature, too. We cannot introduce a wedge between ourselves and nature, otherwise we are bound to repeat the Kantian move of assuming a priori factors that cannot be accounted for from a naturalistic point of view. It follows, then, that (3) there are no a priori elements in our mind, where by a priori elements I mean some unexplained factors which would allow human beings—and *only* them—to play a creative role in the construction of nature itself. The best thing to do, at this point, is (4) to conceive of our conceptual schemes as physical processes that *reflect* reality to some extent. And (5) this means, in turn, that our image of reality is not something separate and independent of reality as such; therefore, there is a *continuum* between our conceptual scheme and reality, in the sense that our conceptual scheme is just one of the forms through which reality happens *to be*. It follows (6) that the realism/anti-realism and idealism/materialism debates, as they usually took place in philosophy up to the present days, make little sense. In fact we are just part of a complex physical process, and there is no discontinuity between nature and our mind which is just a part of it. But (7) at this point the Kantian distinction between *phenomena* and *noumena* makes little sense, too. Since there is only one nature, we are also bound to admit that there is only one ontology. Then (8) Davidson's denial of the scheme/content distinction is vindicated.

The question to be asked is, obviously, the following: *What* are the consequences of this sketchy trend of thought? Note that my previous considerations by no means imply an endorsement of a Davidsonian stance: what I am saying is that thought, language, and mind are just parts of nature. So Wittgenstein is right when he says that some questions cannot be asked because they do not even make sense, but in my view we may interpret him in a way different from the traditional ones that have been thus far put forward. We may well accept Wittgenstein's statement that the existence of the world, for instance, cannot be questioned meaningfully. But this means, in turn, that the linguistic games cannot go on forever. Sooner or later we run into a "hard rock" which is ultimately nonlinguistic and whose existence is the original fact from which everything else stems, including language, linguistic games, conceptual schemes, social practices, and whatever else you care to name. Everything, in sum, can be questioned, except nature. And when someone *does* question it, like the pupil mentioned in *On Certainty*,[16] who will not let anything be explained to him by his teacher, for he continuously interrupts him with doubts concerning the existence of things, we are somehow forced to answer his questions as Wittgenstein's teacher does: "Stop interrupting me and do as I tell you. So far your doubts don't make sense at all."

Obviously, I am aware of the fact that neither the orthodox Wittgensteinians nor Rorty are likely to accept what I have said thus far. But it should also be clear that my unfaithfulness to the orthodox gloss of Wittgenstein's texts is here far less important than the basic point I want to make. And this basic point may be summarized as follows. In order for a picture of the world to exist, there must be something to be pictured, and in order for a reference framework to exist, there must be something this framework matches. Otherwise, we are somehow suspended in a void in which—as Rorty claims—the fact that a sentence has sense depends upon whether another sentence is true, this sentence being *only* about "the social practices of the people who use the marks and the noises which are the components of the sentence." Other thinkers have in turn claimed that only language, conceived of as a set of social practices, guarantees truth. It seems to me that these are just elegant attempts to skip philosophical problems which arise spontaneously in any kind of social practice, and on such an attempt, in fact, Rorty bases his popular thesis of the alleged "end of philosophy."

But things, of course, cannot be that simple. Today we see that many metaphysical issues with which most philosophers can no longer cope are, instead, taken up by practicing scientists. They often express their astonishment when they realize to what extent contemporary philosophy has become "minimalistic," in the sense of being concerned only with small and technical details regarding linguistic usage. Is it Wittgenstein's fault, as the physicist Stephen Hawking has recently suggested? The answer is yes if we stick to Richard Rorty's interpretation of Wittgenstein. The answer is instead no if we are bold enough to overcome the orthodox gloss that has progressively transformed Ludwig Wittgenstein into a kind of mausoleum mummy. I would like to add that Rorty is certainly right when he claims that "the linguistic turn was a second attempt to find a domain which would overarch those of the other professors."[17] The time has clearly come to say: All right, the linguistic turn has been a wonderful adventure, but now it has become outdated: let us try something else. The problem—we all know—is that right now there does not seem to be anything else available (except for Heidegger and hermeneutics), so that philosophy's survival is in danger. But this should not prevent all honest philosophers from recognizing that the time of the linguistic fairy tales is over.

4. WEINBERG'S DREAMS

Stephen Hawking's well-known hypothesis of the unbound universe can be expressed in the following way:

> A universe which is completely self-contained, having no boundary or edge and no beginning or end. In Hawking's own words the universe "would simply BE."[18]

This is indeed a good example of how contemporary philosophy left to science the task of doing a metaphysics based on observational findings and hypotheses. It is obvious that, if Hawking's idea is correct, the old question, Why does the universe exist? does not even make sense, since being at this point would become self-explanatory. Someone might object that this is not metaphysics, and he is right, if by "metaphysics" we intend, for instance, the systems built by Aristotle or Hegel. The fact is, however, that metaphysics—just like science—changes a great deal with the passing of time. Contemporary philosophers no longer sit down and imagine the structure of reality by using mere thought and pure deductive reasoning: they must—or, at least, they should—take into account scientific results and help scientists to build up a coherent view of reality. Popper's "cosmology," about which we talked previously, fits well into this picture.

It should be clear, in any event, that this *is* metaphysics, because in adopting such a stance scientists are in fact looking for the first principles of being. As we said before, if the expressions "metaphysics," "being," "first principles," and so on appear for some reason outdated and bothersome, we may well replace them with new terms, as Popper does by defining "cosmology" as a research that appears to be plainly metaphysical in character.[19] But, no doubt, this is only a linguistic device that cannot hide the real nature of the problem at stake. Note that what we have just said makes Hawking's criticism of the linguistic turn in contemporary philosophy[20] even more understandable. For his part, Steven Weinberg gives us a fair picture of both scientists' basic realistic attitude and their ongoing search for the first principles:

> Sitting at my desk or at some café table, I manipulate mathematical expressions and feel like Faust playing with his pentagrams before Mephistopheles arrives. Every once in a while mathematical abstractions, experimental data, and physical intuition come together in a definite theory about particles and forces and symmetries. And every once in an even longer while the theory turns out to be right; sometimes experiments show that nature really does behave the way the theory says it ought. . . . Our present theories are of only limited validity, still tentative and incomplete. But behind them now and then we catch glimpses of a final

> theory, one that would be of unlimited validity and entirely satisfying in its completeness and consistency. We search for universal truths about nature, and, when we find them, we attempt to explain them by showing how they can be deduced from deeper truths. *Think of the space of scientific principles as being filled with arrows, pointing toward each principle and away from the others by which it is explained. These arrows of explanation have already revealed a remarkable pattern: they do not form separate disconnected clumps, representing independent sciences, and they do not wander aimlessly—rather they are all connected, and if followed backward they all seem to flow from a common starting point. This starting point, to which all explanations may be traced, is what I mean by a final theory.*[21]

This is a basic realistic attitude because we gather from it that the scientist keeps questioning nature, and never believes that it can be identified with his mathematical abstractions and physical intuitions. Nature, in turn, gives answers to his inquiries, and the answers may be positive, negative, or simply interlocutory. Unlike the analytic philosopher, though, he never has the feeling of talking just—and only—to himself. On the other hand, even the metaphysical search for the first principles is rather clear, provided we read Weinberg's words without philosophical prejudices. What other sense can we make of those arrows of explanation that all seem to flow from a common starting point?

And so we are back, once again, to the problem of the *attitude* towards nature, a problem which is the main source of the current gap between operative science and the vast majority of contemporary philosophy, a fact we have already mentioned several times in the course of this work. As it was remarked before, modern and contemporary philosophy are, with a very few exceptions, anthropocentric. They are speculative systems centered almost exclusively on human beings and their perspective on the world: humanity is no longer seen as a component, however important, of nature, but rather as a wholly autonomous entity that has nothing to do with nature itself or, even worse, as an entity which is bound to set itself *against* nature. No wonder, then, that nature does no longer seem to be a worthy subject of inquiry. As we said previously, this is not an exclusive feature of contemporary analytic thought. The most powerful expression of the basic antinaturalistic attitude of modern philosophy may actually be found in the poetic vein of Friedrich Nietzsche:

> Great star! What would your happiness be, if you had not those for whom you shine! You have come up here to my cave for ten years: you would have grown weary of your light and of this journey, without me, my eagle and my serpent.[22]

Nietzsche's Zarathustra, who deems the Sun useless and meaningless without the presence of human beings, is both the perfect characterization of the anthropocentric stance of modern philosophy and the poetic and messianic expression of all those trends of thought that endorse absolute anthropocentrism because, allegedly, human beings are never able to take into account nature-in-itself, but only nature-for-us. Kant, idealism, and analytic philosophy all claim that, since humans always need the language/thought to deal with nature, they somehow build up reality (in the sense of creating it).

What we find in the words of a professional physicist like Weinberg, instead, is a constant search for the fundamental principles which can explain why nature is the way it is. Today—he states—we believe that atoms behave the way they do in chemical reactions because the physical principles that govern the electrons and electric forces inside atoms leave no freedom for the atoms to behave in any other way. What Weinbergs tells us, in sum, is that scientists are pursuing the discovery of explanations built *into* the structure of nature itself. Note how deep and complete is the reversal of the linguistic turn brought in by logical positivism and the analytic tradition. What scientists do is thus a new kind of metaphysics supported by empirical inquiry and operational practice: it is nothing but the old dream to find out what the first principles of being are. There is even a reevaluation of the pre-Socratic thinkers, since Weinberg and many other scientists, after all, feel themselves more in tune with them than with modern philosophers, although realizing that the pre-Socratics overlooked the quantitative dimension of phenomena (and, obviously, this is just a shortcoming due to the changed historic context).

But the story is by no means ended. Weinberg, in fact, openly talks of "a necessity built into nature itself,"[23] deeming that what science is about is just the discovery of explanations which stem from the structure of nature. Clearly his approach is opposed to a line of thought—now very popular within philosophy of science—according to which the fundamental character of a lawful proposition is not the aspect of being-a-law but, rather, the *relational* aspect of being-maintained-as-a-law by the knowing subject. And it means, too, that we should abandon the view that the dogmas of empiricism are those about which Quine talks in his famous paper "Two Dogmas of Empiricism." There is, instead, only one, very basic dogma: that brought in by the linguistic turn, namely, the idea that philosophy must adopt some type of *semantic ascent* in order to cope with reality. Beneath that dogma lies Kant's conception of space and time as a priori faculties of human intellect.

As was previously hinted, we need now a *semantic descent* capable of rebuilding the broken bridges between philosophy of science and operative science. It is in any event obvious that the damages brought in by the linguistic turn depend on the conception of language we adopt because, despite what the orthodox analytic philosophers say, there is more than one conception of language at out disposal. Did anyone really succeed in showing that the view of language as an a priori factor is the sole and correct one? I do not think so. Naturally, if you take the above view to be a truth of faith, and then deduce from it the consequences you have already in mind, everything appears to be all right. But it is also clear that, should we question this truth of faith, showing that it has a lot of shortcomings, then the conception on which the linguistic turn rests falls apart altogether. Since Quine and most analytic thinkers always talk of "scientific philosophy," "scientifically oriented mind," "contiguity between science and ontology," "empirically testable procedures," and so on, let us really be scientifically and empirically oriented philosophers by taking seriously the findings of natural science.

The logical positivists cleverly realized that modern science has pushed philosophy into a corner from which it is difficult to come out safely. So, they reacted by finding a job specifically devised for philosophers (i.e., analysis of language), and afterwards they proceeded to transform linguistic analysis into a new metaphysics which is supposed to provide philosophers with *normative* rules (obviously set forth by the logical positivists themselves). At this point philosophy was certainly saved from extinction, but quite a high price—at least from a purely neopositivist viewpoint—had to be paid: that of adopting an unconscious, although very fascinating, new metaphysics. The problem to be faced is thus the following: If we show that linguistic analysis is not *that* important, what is the philosopher supposed to do? Is there a risk, so to speak, of a sudden "philosophical unemployment"?

For example, after having intelligently realized that the linguistic turn was a rather desperate attempt to keep philosophy an armchair discipline, Richard Rorty invokes a sort of dissolution of philosophy into general culture, so that the philosopher may become an amateur who speaks of everything without taking anything seriously. This is no joke at all: it is, rather, a serious matter. It might be argued that the fall of logical empiricism is also the fall of the last possible defense that twentieth-century philosophy set up against scientific imperialism, and now that scientists are more or less telling us "Be philosophers again!" we find ourselves in a big quandary. Weinberg, for instance, gives us a picture that equates current physical research with an endless chain of "why?"—not just "how?" but "why?"—which is the opposite of the neopositivist picture of scientific enterprise.

He knows, of course, that today philosophers no longer like the word *why*; but, on the other hand, he does not seem to be too impressed with this finding.

Weinberg does not take the anti-why stance of logical empiricism and analytic philosophy seriously while, in turn, he is not afraid of coping with ontological and metaphysical problems, given also the fact that most likely, according to his view, it is the practicing scientist—and not the philosopher—who can actually answer the why-questions. So we are back to our old problem: what is the philosopher supposed to do given this situation, and following the bitter discovery that linguistic analysis is not so central after all? In my opinion, we are bound to recognize that science is entitled to exert some type of imperialism on such traditionally philosophical fields as metaphysics, ontology and the theory of knowledge, while in ethics, philosophy of politics, and other fields this imperialism will be less direct and more mediated.

In any event, a very important fact should at this point be noted. As we hinted at in other parts of our work, philosophers are not allowed to reject this state of affairs because it may lead to their unemployment: love for truth should be deemed more important than any academic status, be it philosophical or not. Both metaphysics—which, as we saw, has not been eliminated at all—and the theory of knowledge must be reshaped and rethought in order to make them not only compatible with, but also contiguous to, operative science. On the other hand, philosophy of science, as I argue in the final section of this book, has nothing else to do but adhere to naturalism. It cannot ignore what practicing scientists are actually doing, neither is it allowed to cultivate impossible dreams of independence. Philosophers of science can occupy all those spaces that scientists leave empty because they are too busy with the technical details of their daily work. For instance, they may try to give tentative answers to the many "whys" scientific activity keeps raising and, given the inevitable parasitic (in a good sense of the term) character of their discipline, philosophers of science may even try to build comprehensive and consistent outlooks on natural reality. But the validity and consistency of such outlooks always need to be verified with scientists themselves. No doubt many will find this picture disappointing, if compared to the bold constructions envisioned by the logical positivists, whose final objective was that of equating the philosopher of science to a superscientist endowed with normative power. There is, however, no other way out, and a modest (but useful) task is better than an important (but pointless) one.

Note that this modesty of intents is likely to eliminate many misunderstandings that currently divide philosophers of science from practicing scientists. As Weinberg has it:

> Talk of more fundamental truths makes philosophers nervous.[24] We can say that the more fundamental truths are those that are in some sense more comprehensive, but about this, too, it is difficult to be precise. But scientists would be in a bad way if they had to limit themselves to notions that had been satisfactorily formulated by philosophers. No working physicist doubts that Newton's laws are more fundamental than Kepler's or that Einstein's theory of photons is more fundamental than Planck's theory of heat radiation. . . . Ludwig Wittgenstein, denying even the possibility of explaining any fact on the basis of any other fact, warned that "at the basis of the whole modern view of the world lies the illusion that the so-called laws of nature are the explanations of natural phenomena." Such warnings leave me cold. To tell a physicist that the laws of nature are not explanations of natural phenomena is like telling a tiger stalking prey that all flesh is grass. The fact that we scientists do not know how to state in a way that philosophers would approve what it is that we are doing in searching for scientific explanations does not mean that we are not doing something worthwhile. We could use help from professional philosophers in understanding

> what it is that we are doing, but with or without their help we shall keep at it.[25]

This is rather disquieting for philosophy of science as it is currently practiced, because it is tantamount to saying: "Well, we do not care at all about your antimetaphysical fears or your antipragmatic love for precisely defined words." How can philosophers of science deem positions like this irrelevant to their own discipline? Since analytic philosophers almost always think that details are more important than comprehensive pictures, some of them might object that Wittgenstein was not a philosopher of science, thus dismissing Weinberg's criticisms. But this objection has no value, because the Austrian-born thinker has had—through the linguistic turn—an extremely deep influence on the philosophy of science of our century.

On the other hand, as we remarked in the first chapter, the logical positivists are not—as many post-empiricist philosophers of science like to say—totally wrong. Weinberg, along with a high number of contemporary physicists, agrees with them on the theme of *reductionism*, a stance he openly endorses, not out of a renewed primacy of physics on any other natural science, but because of its capacity to ascertain the fundamental laws of nature. During the late nineteenth century and the early part of the twentieth—he argues—many scientists seem to give up on the old dream of Newton that chemistry and all other sciences would be understood in terms of physical forces; for them, chemistry and physics had become coequal sciences, each separately near completion, and "to whatever extent there was a widespread sense of completeness in late-nineteenth-century science, it represented only the complacency that comes with diminished ambition."[26] This, in turn, is nothing but the neopositivist project of a unified science with physics in a position of leadership, that is, the subject of the famous *Encyclopaedia of Unified Science*. Not only is this so: it means that the neopositivists, no matter how many shortcomings we are able to find in their philosophy of science, got the essential thing right, after all. It is evident that this strategy, if it is justified, would change the whole picture of philosophy of science which has become predominant in the last few decades, following the post-empiricist turn. We can thus find precious insights in both logical positivism and post-empiricism, although neither of the two movements has been able to give us a totally faithful picture of current scientific practice. It may well be, as Peter Strawson has recently remarked, that:

> There is, as it were, a tendency to intellectual imperialism on the part of, or on behalf of, different disciplines—so that now, say, physics, now biology, now psychology, now economics or sociology or anthropology—or even linguistics—will be felt to hold the master-key to general understanding; so that everything is to be understood in the light of physics, or biology, or the processes of economic production—or whatever else it may be. It goes almost without saying that any theory which claims to offer a general picture of reality, with everything in its place—but which is in fact constructed under the dominance of some such particular interest—that any such theory will be likely, indeed certain, to involve exaggeration and distortion.[27]

It might be noted, however, that such statements do not seem to apply to our previous reasoning, which involves taking into account the fundamental laws of nature on which everything else is based. What is wrong in assuming that some truths are less fundamental than others to which they may be reduced, as Weinberg does when he claims that chemistry can be ultimately reduced to physics? And why should the research for "first principles," "fundamental laws," or "ultimate constituents" be wrong? A tentative answer may be the following: because this means getting involved in *metaphysics*. It is, as we have already noted, a new type of metaphysics that always needs the proximity to science but, still, its objective is to give comprehensive answers. Now, I do not think this is a wrong kind of "reductionism." It is, rather, the natural human desire to explain something complex by means of its constitutive parts. In this sense, reductionism is nothing but the perception that there are deeper scientific principles, which may in turn be traced back to one simple connected set of laws. In this sense, endorsing reductionism means simply to comply with the order of nature.

Weinberg's ontological convictions are well expressed by the following statements: "It had at last become clear that chemical phenomena are the way they are because of the electrical interactions of electrons and atomic nuclei. . . . The success of quantum mechanics in calculating the properties of very simple molecules made it clear that chemistry works the way it does because of the laws of physics."[28] So there is, according to him, only one basic reality, whose deepest comprehension is in turn founded on the understanding of the fundamental laws of physics; and he goes even further, claiming that ultimately all those laws can be reduced to an extremely limited set of laws. Weinberg knows even too well that any talk of a final theory is likely to disturb most philosophers: those who talk of a final theory will be accused of reductionism or even physical imperialism. But this is due to a current misunderstanding, since many people are led to conclude that a final theory in physics would mark the "end of science."

It may be answered, however, that a final theory would not end scientific research, because many phenomena, among which is thought itself, will still need explanation, whatever final theory turns out to be discovered. A final theory, thus, will be final in only one sense: it will bring to an end the ancient search for those principles that cannot be explained in terms of deeper principles. As was previously noted, Weinberg talks of tracing the arrows of explanation back toward their source, so that we will eventually be able to discover a "striking convergent pattern" which is likely to be the deepest thing we have yet learned about the universe. Taking, for example, a piece of chalk, we have discovered the *causal* chains that explain the presence of chalk today in the world, and this seems to match Wesley Salmon's picture, which attributes a primary importance to the causal structure of reality.[29]

On the other hand, as was hinted before, the old materialistic view of the world is no longer in good shape. Following the fusion of relativity theory with quantum mechanics, we now have a new view of reality, and precisely one in which matter has lost its central role. The Newtonian theory which says that matter consists of tiny immutable particles acting one on another through forces is, according to Weinberg, "a dead end." Obviously this has enormous bearing on the realism/anti-realism debate, since fields, and not particles, are taken to be the ultimate components of reality:

> The equations of a field theory like the standard model deal not with particles but with fields; the particles appear as manifestations of these fields. The reason that ordinary matter is composed of electrons, protons, and neutrons is simply that all the other massive particles are violently unstable. The standard model qualifies as an explanation because it is not merely what computer hackers call a kludge, an assortment of odds and ends thrown together in whatever way works. Rather, the structure

> of the standard model is largely fixed once one specifies the menu of fields that it should contain and the general principles (like the principles of relativity and quantum mechanics) that govern their interactions.[30]

The striking bearing of these remarks on ontology can hardly be denied. Because it is clear that *if* fields—and not particles—are the basic constituents of reality, *then* it is necessary to start off the current realism/anti-realism debates on new bases. Since we may say that something is explained even where we have no assurance that we will ever be able to deduce it, we can conclude that scientific explanation turns out to be largely a matter of intuition for practicing scientists. If they see that the explanation *works*, then everything is all right, and they do not even bother to find out a precise line of deduction. The "because" that scientists so often use in their explanations has nothing to do with our ability actually to deduce anything, and the fact that *there is* an order of nature which does not depend on our conceptual abilities is assumed from the beginning. Why? Because otherwise, as we already remarked in the preceding chapters, there would be no scientific practice at all.

Weinberg's work, thus, is not a refusal of philosophy per se, but rather a refusal of the image of science that twentieth-century linguistically oriented philosophy of science has put forward. On the one hand, analytic philosophy of science does not explain at all the dynamic aspect of scientific enterprise, while on the other, the post-empiricists dangerously dissolve science into history and sociology. We have in sum opposed—but connected—errors, and it should be noted that the recent and greatly popular reduction of science to social practice does not work either. Sociology of science attacks scientific objectivity by claiming that science is simply a matter of social construction, a position that Weinberg bluntly rejects noting that "it is simply a logical fallacy to go from the observation that science is a social process to the conclusion that the final product, our scientific theories, is what it is because of the social and historical forces acting in this process."[31]

As a matter of fact, this radical attack on the objectivity of scientific knowledge can be traced back to the old positivist refusal to talk about anything that is not directly observed: if one starts with such an unwarranted assumption, it is quite obvious that neither quantum field theories nor objective laws of nature can be taken seriously. But I would like to add that there is even more to say. First, the aforementioned old positivist assumption is just the basis of the contemporary refusal of metaphysics conceived of as a search for the first principles of reality supported by the findings of scientific practice. Alfred J. Ayer's attack on metaphysics, as it is envisioned in *Language, Truth and Logic*, could not have been carried out without the presence of this positivist dogma. Second, probably not even Weinberg takes into sufficient account the enormous influence that the linguistic turn has had on the philosophy of science of our century. The reduction of reality to language, which in turn is so much indebted to Kant's conception of a priori knowledge, is the second root of the problems that philosophy of science is facing nowadays. These two basic assumptions—(i) the positivist refusal to talk about anything that is not directly observed, and (ii) the analytic reduction of reality to language—form what I would like to call an "explosive mixture." W. V. O. Quine's work, as we showed in chapter 2, is a good example of the effects this mixture may have on ontology, philosophy of science, and the theory of knowledge.

In any event, neither Weinberg nor other practicing scientists seem to be worried by the quibbles currently thriving in philosophical circles (and particularly within philosophy of science). He says, in fact, that "These radical critics of science seem to be having little or no effect on the scientists themselves. I do not know of any working scientist who takes them seriously. The danger they present to science comes from their possible influence on those who have not shared in the work of science but on whom we depend, especially on those in charge of funding science and on new generations of potential scientists."[32]

We thus have a practical danger, and not a theoretical one, even though the practical nature of the mounting antiscientific spirit must not be underestimated. In the first chapter of this book we saw, for instance, that the diffusion of the antiscientific spirit in contemporary Western societies has brought in irrationalism, which seemed definitively defeated at the end of the second World War. Not surprisingly, Weinberg evokes the name of Oswald Spengler when talking of the radical attack on science as a symptom of a broader hostility to Western civilization as such.

On the other hand, the critics of Western civilization, as we noted in the first chapter dealing with Feyerabend's positions, often claim that since other regions of the world have maintained their own languages and religions, so they could also have kept their own scientific traditions, developing scientific laws different from ours. But it is easy to verify that this is not the case, and Weinberg points out that: "The physics of modern Japan and India is the same as that of Europe and America. . . . The experience of listening to a discussion of quantum field theory or weak interactions in a seminar room in Tsukuba or Bombay gives me a powerful impression that the laws of physics have an existence of their own."[33]

Are we allowed to conclude that Weinberg—along with Hawking and many other famous scientists nowadays—is right in criticizing the way philosophy of science has developed in the twentieth century? It should be clear by now that my answer to such a question is plainly positive. Very often philosophers of science do not seem to understand what science is really about; for example, they seem very worried by the presence of experimental anomalies, to which Weinberg replies that any theory like Newton's theory of gravitation is always plagued by experimental anomalies. There is no theory that is not contradicted by some experiment.[34]

These statements sound weird to the ears of many philosophers of science, but they simply reflect the fact that if some experiment contradicts the theory, practicing scientists will not therefore abandon it. And this confirms the view, already hinted at in our work, that the philosophers of science should check more closely what practicing scientists are actually doing. Weinberg also criticizes the "obscure jargon" so often adopted by philosophers of science, quoting a book published in 1984.[35] It is a book—so our physicist claims—written in a jargon so impenetrable that, most likely, it aimed at impressing those who confound obscurity with profundity.[36] Philosophy of science books, as a matter of fact, are often dominated by a useless formalism, and this happens because so many philosophers nowadays believe that formalism is the best way for getting a deeper comprehension of science itself. But, as Weinberg's statements clearly show, practicing scientists do not recognize themselves in that formalism; they rather ask philosophers to help them in clarifying real issues, and not to compete with them at the formal level. The opposite approach, as we know even too well, ends up frequently in telling jokes. In any case, both the formal and the historical-sociological approaches in philosophy of science have by and large lost the contacts with *concrete* science. They both grew according to *internal* standards, which in turn do not permit taking into account the fundamental problem of rebuilding suitable bridges between science on the one hand, and philosophy of science on the other.

The very important thing to understand is, eventually, the following: Any failure or shortcoming in scientific practice is not likely to change the basic realist attitude of scientists. Two main reasons may explain this outcome: (1) because otherwise it would be impossible to pursue any meaningful scientific work, and (2) because in any event there is a reality to catch out there, and the fact that we do not yet comprehend it in a complete manner is no good reason for adopting an anti-realist stance. Scientists' research is about *nature itself* because they feel that just this kind of research can explain why the world is the way it is: "The study of quarks and electrons is fundamental not because all ordinary matter is composed of quarks and electrons but because we think that by studying quarks and electrons we will learn something about the *principles* that govern everything."[37]

5. A NATURALISTIC PHILOSOPHY OF SCIENCE

Our language and conceptual schemes certainly play an important role in the process of scientific enterprise, but this role is not as exclusive as the linguistically oriented philosophers of science seem to believe. This may be a consequence of the presumption concealed in many anthropocentric visions of modern philosophy: in this sense, I think that the analytic tradition is one of the main heirs of this presumption. We are—as science tells us—just a small episode in the history of reality, and should not thus think that reality is meaningful only when we get to know it: scientists do not need Michael Crichton's *Jurassic Park* in order to know that human history is a part of the (much longer) history of the universe. Naturally, as we have seen before, one can always reply: "Why should I care about a world without mankind?" I find this answer totally misleading. We must change our vision of the world, trying to think objectively, since this is what science always does, and philosophy of science must take this point of view into account just to understand what science is all about. For this reason, instead of talking of a "realist" or an "anti-realist" philosophy of science, I believe we should talk of a *naturalistic* philosophy of science.

A possible objection to what I have claimed might be: "After all, you do not take into sufficient account the sophisticated analyses made by many analytic philosophers." This may to a certain extent be true, but I made this choice because I am convinced of the following fact. The philosopher of language (including the linguistically oriented philosopher of science) is like a spider who spins its web to capture insects. If you use his instruments too frequently, sooner or later you will be caught by the web. So we must instead try to rip the web off, abandoning a purely linguistically oriented philosophy of science. We must break the threads of this web just to get an objective vision of reality. It is extremely difficult, I know, because we are constantly limited by our subjectivity, but nevertheless we must try.

We have already noted before that there is a strict parallelism between a large part of contemporary linguistically oriented philosophy and idealism, and now I want to say something more on this topic. The idealist claims:

a. we cannot step out of *thought*: if we admit that there is an external reality which transcends thought, then, by the same act of thinking it, this alleged external reality is no longer transcendental. It follows that we can never overcome the cognitive identity between being on the one side, and thought on the other.

But most analytic philosophers, including the linguistically oriented philosophers of science, would paraphrase (a) in the following manner:

b. we cannot step out of *language*: if we admit that there is an external reality which transcends language, then, by the same act of talking about it, this alleged external reality is no longer transcendental. It follows that we can never overcome the cognitive identity between being on the one side and language on the other.

As was mentioned previously,[38] we have to bear in mind that language is in the analytic tradition—and in contemporary hermeneutics as well—something much more important than the ordinary language we use daily to communicate with other people and to refer to an external reality. Language becomes, more or less, the logical structure of the world, an extremely powerful and a priori factor which—borrowing a distinction introduced by Wilfrid Sellars[39]—includes both overt (linguistic) and covert (mental) semantic categories; in this sense language just becomes a variant—although with some original characters—of classical idealism's thought.

We can trace the development of this view back to the classics of analytic philosophy—Gottlob Frege, Ludwig Wittgenstein in the *Tractatus Logico-Philosophicus*, Rudolf Carnap, and many others. Philip Kitcher has recently claimed, in this regard:

> Frege's investigations are commonly viewed as a decisive turn, one that dethroned epistemology from its central position among the philosophical disciplines and that set the philosophy of language in its place. In retrospect, we can trace a great lineage from Frege, leading through Russell, Wittgenstein, and Carnap to the professional philosophy practiced in Britain, North America, Australasia and Scandinavia in the postwar years. . . . Frege's opposition to what he perceived as intrusions from psychology or biology is evident from celebrated passages in the *Grundlagen*. The methodological stance he inspired becomes explicit in propositions of the *Tractatus*. . . . Anglo-American philosophers have explored a wide range of disciplines, using ideas from psychology, biology, political science, economics, and the arts to reformulate traditional questions in epistemology and metaphysics. Some of their endeavors mark the return of epistemological naturalism, scorned by Frege and labeled as illicit philosophy by Wittgenstein.[40]

But let us note that even the later Wittgenstein, notwithstanding the opinion of many authoritative scholars, is not that distant from this conception. Think of his linguistic games, a notion through which reality is bound to the very limits of our language. This conception is exactly what Gustav Bergmann has named the "metaphysics of logical positivism," a metaphysics which is basically shared by the analysts of ordinary language. Kitcher's essay, however, although it is very interesting, has some shortcomings too. For example, he claims that Quine, thanks to his naturalized epistemology, endorses naturalism. As previously said, instead, I think that Quine holds an antinaturalist approach because of his basically neopositivist conception of language. Furthermore,

Kitcher is interested in epistemological naturalism, while I think that naturalism must be based on ontological premises.

The key operation, in this and other cases as well, is to try to get rid—as much as we can—of our subjectivity, which erroneously suggests that reality exists as long as is known to us or has some kind of link with us. We may indeed sit down, close our eyes, and perfectly imagine a situation in which mankind is not present and reality takes care of itself. The picture we get is astonishingly clear, provided we are honest enough to see it: a world that develops without the presence of human beings. And, of course, we can also find lots of evidence supporting this picture: as I said before, science gives us many examples of worlds developing without anything similar to human life. For this reason I deem important the examples we previously gave concerning paleontology and astronomy.

It is worth noting, however, that if I carry on the opposite experiment, trying to see men without a world surrounding them, it will not work. So the situation is clear enough: I can imagine a world without mankind, but no men without a world. It is not our language or conceptual schemes that have the ontological primacy, but rather a reality which is independent of both human language and conceptual schemes. This in turn means that when we read—maybe in a daily newspaper—that "scientists ask again and again why nature is the way it is," we should take this statement seriously because philosophy of science heavily relies on those scientists' basic realist attitude. After all, there is no reason to assume that there are *noumena*, that is, objects unknowable even to science: this is an unwarranted metaphysical assumption. There is no reason to believe, in other words, that in a table—or in an electron—there is something besides what science tells us. This is an old philosophical vice, and the time has come to get rid of it. If someone objects that our science, however, has not yet been able to explain completely the structure of reality, we can underline, once again, the *not yet*: considering what happened in the past few centuries, there is indeed no motive to presume that scientific research will stop its successful journey.

We live in a century in which philosophy has been largely dominated by preoccupation with language and linguistic analysis. As I said before, not only logical empiricism and analytic philosophy of ordinary language took this stance, but also hermeneutics, whose representatives think that the only job philosophy can perform is interpreting "cultural texts." And Wittgenstein himself is not that distant from this trend of thought, particularly when he claims that his philosophical method switches all questions on truth into questions of meaning.

So it seems to me that most of contemporary philosophy became just "obsessed" with language (which is different from being seriously interested in it), forgetting that linguistic analysis is just one philosophical method out of many. The current opinion is that we cannot step out of language, because either (i) language and reality are one and the same thing, or (ii) language is the only thing we can know, reality in itself being unknowable. But, besides all that, even man himself has been identified with language, just forgetting the fact that man came first and language later. Language is, in fact, recent, as science shows us, and many parts of our personality are guided by nonlinguistic criteria. This is the proof of what I said before: we are just a small nonlinguistic part of a nonlinguistic reality. Language is *not* reality, but just a product of the human mind created essentially for practical and social purposes. It can indeed be said that, man being a small part of reality, and language being a limited part of man, language itself is a subpart of a subpart of reality. From this one can easily gather how absurd it is to believe that reality and language are one and the same thing.

All this has important consequences on (i) the philosophy of science, (ii) philosophy itself, and (iii) the entire philosophical vision of the world that surrounds us (although I do not think that scientists need to be told: they already know it). As language is a relatively new entry in the history of reality, it cannot have any sort of ontological supremacy. Not only is this so: it is likely to hide the nonlinguistic dimensions of our human nature while, being restricted to mankind, it cannot explain an enormously large number of the features of reality as such. Let us then stress that science, instead, always tries to enlarge (and to deepen) as much as possible our vision of reality and, in order to do this, we must push our sight both towards the past—when mankind did not yet exist—and the future—when mankind perhaps will no longer be there. This in turn means trying to get a good comprehension of *reality as a whole*: human and nonhuman, mental and nonmental, linguistic and nonlinguistic. Certainly language will have a role in this enterprise, although not a unique one.

A philosophy of science which takes language to be the only feature of reality that deserves to be investigated is then, intrinsically, a form of reductionism which, unlike the one endorsed by Weinberg and other scientists, is unwarranted. Trying to reduce an extremely complex reality to something much simpler, this approach cannot even explain *why* language was born and *for what purposes*. Language becomes a sort of divinity which is supposed to explain everything while, since it is a rather mundane and imperfect product of the human mind, it needs indeed to be explained by tracing its origins, which, as we have noted, are both social and practical. So the authors who—like Quine—mean to replace any argument on reality with arguments on human language that talks about reality are bound to miss, in my view, the richness of reality itself: this is the reason why we need a *semantic descent* replacing Quine's semantic ascent.

Science, through a historic and evolutionary process, shows that, step by step, it is possible to penetrate the structure of reality, and this in turn means that—thanks to science again—our comprehension of "what there is" gets better and better. It is like a big puzzle formed by an incredibly high number of components: we do not know if mankind will ever be able to achieve a complete comprehension of the picture, but on the other hand we know that science marches in that direction. Only in this context does Quine's notion of contiguity between ontology and science make sense. Thus the possibility that mankind will never get a *complete* picture is no real deficiency, provided we recall the preceding considerations on the absence of perfection in the world.

NOTES

CHAPTER 1

1. See G. D. Romanos, *Quine and Analytic Philosophy: The Language of Language* (Cambridge: MIT Press, 1984), in which the author argues instead in favor of Quine's independence of both logical positivism and the analytic tradition.

2. I. Hacking, *Representing and Intervening: Introductory Topics in the Philosophy of Natural Science* (Cambridge: Cambridge University Press, 1993), 42–45.

3. R. Carnap, "The Elimination of Metaphysics through Logical Analysis of Language," in *Logical Positivism*, ed. A. J. Ayer (New York: The Free Press, 1959), 60–81. The original German version, "Überwindung der Metaphysik durch Logische Analyse der Sprache," was published in *Erkennis* 2 (1932): 219–41.

4. R. Rorty, "Wittgenstein, Heidegger, and the Reification of Language," in *The Cambridge Companion to Heidegger*, ed. C. B. Guignon (Cambridge: Cambridge University Press, 1993), 337–57. The article originally appeared in Rorty's *Essays on Heidegger and Others* (Cambridge: Cambridge University Press, 1990).

5. See chapter 5, sections 1 and 2.

6. This is the case, for instance, with the physicist Stephen Hawking, as we shall see later on in chapter 2, section 1.

7. See note 3, above.

8. Carnap, "Elimination of Metaphysics," 67.

9. A. J. Ayer, *Language, Truth and Logic* (London: Penguin Books, 1990), 32–33. As is well known, the book was first published in 1936.

10. L. Wittgenstein, *Tractatus Logico-Philosophicus*, ed. A. J. Ayer (London: Routledge and Kegan Paul, 1961), 4.113–115, p. 49.

11. Ibid., 4.112, p. 49.

12. See chapter 2, section 4.

13. W. V. O. Quine, "Identity, Ostension, and Hypostasis," in *From a Logical Point of View* (New York: Harper and Row, 1953), 79.

14. R. Carnap, *Meaning and Necessity* (Chicago: University of Chicago Press, 1947). This theme was developed by Polish thinkers like T. Kotarbinski and K. Adjukiewicz. See, for example, K. Szaniawski, ed.,

The Vienna Circle and the Lvov-Warsaw School (Dordrecht: Kluwer, 1989); and M. Marsonet, *Linguaggio e conoscenza: Saggio su K. Adjukiewicz* (Milan: Angeli, 1986).

15. See especially Quine's famous essay, "Two Dogmas of Empiricism," in *From a Logical Point of View*, 20–46.

16. See another famous essay by Quine, "On What There Is," in *From a Logical Point of View*, 1–19. A clear example of analytic metaphysics is given in the well-known book by A. Plantinga, *The Nature of Necessity* (Oxford: Clarendon Press, 1974).

17. A good example is D. Oldroyd's *The Arch of Knowledge: An Introductory Study of the History, Philosophy, and Methodology of Science* (New York: Methuen, 1986).

18. H. Putnam, *Il Pragmatismo: una questione aperta* ("Pragmatism: an open question") (Rome and Bari: Laterza, 1992), 24. As far as I know, this book, which is made up of three lectures given by Putnam at the University of Rome, has no English version, and the present English translation is mine. The quotation is drawn from W. James's essay, "Spencer's Definition of Mind as Correspondence," in his *Essays in Philosophy* (Cambridge: Harvard University Press, 1978), 21.

19. David Oldroyd, in the book mentioned in note 17, shows that the creation of the relativity theory by Albert Einstein cannot be accounted for by using a neopositivist strategy of explanation.

20. K. R. Popper, *The Logic of Scientific Discovery*, rev. ed. (London: Hutchinson, 1968), 15–19.

21. As we shall verify later on, Popper's position is compatible with a certain revival of metaphysical attitudes currently thriving among practicing scientists.

22. Hacking, *Representing and Intervening*, 43.

23. P. K. Feyerabend, *Three Dialogues on Knowledge* (Oxford: Blackwell, 1991). The present quotation is drawn from the Italian translation, *Dialoghi sulla conoscenza* (Rome and Bari: Laterza, 1991), 94.

24. I am obviously aware of the fact that the later Wittgenstein, especially in such books as *Philosophical Investigations* (New York and London: Macmillan, 1958) and *On Certainty* (Oxford: Blackwell, 1969), adopted a stance that closely resembles the pragmatist positions. Hilary Putnam, for instance, entitled the second chapter of his book, *Il pragmatismo: una questione aperta* (Rome and Bari: Laterza, 1992), "Was Wittgenstein a Pragmatist?" It seems to me, though, that the absolute primacy of language endorsed *even* by the second Wittgenstein somehow lessens the efficacy of these insights. Furthermore, the thought of the second Wittgenstein may lead to conclusions like the following, which I deem misleading: "Objectivity and rationality must be things that we forge for ourselves as we construct a form of collective life. So the work of Copernicus is undone. Human beings are back in the centre

of the picture. . . . The things we had seen ourselves as answerable *to*, we are now answerable *for*" (D. Bloor, *Wittgenstein: A Social Theory of Knowledge* [London: Macmillan, 1983], 3).

25. S. Weinberg, *Dreams of a Final Theory* (New York: Pantheon Books, 1992), 190.

CHAPTER 2

1. See for example the book by Alex Orenstein, *Existence and the Particular Quantifier* (Philadelphia: Temple University Press, 1978), in which the author argues that existence is not a matter of logical quantification; and the article by Nicholas Rescher, "The Equivocality of Existence," in *Studies in Ontology*, American Philosophical Quarterly Monograph Series, no. 12 (Oxford: Blackwell, 1978), 57–66.

2. The reference work in this case is still G. Bergmann's *The Metaphysics of Logical Positivism*, 2d ed. (Madison: University of Wisconsin Press, 1967).

3. The essential referentiality of human language has been pointed out by many representatives of the phenomenological school.

4. S. Hawking, *A Brief History of Time: From the Big Bang to Black Holes* (New York: Bantam Books, 1988).

5. Ibid., 174–75.

6. Weinberg, *Final Theory*, 166–67.

7. Hilary Putnam, for instance, expressed this opinion when talking of Weinberg's theses in a lecture he gave at the Center for Philosophy of Science of the University of Pittsburgh on 3 March 1993. In my opinion, Putnam cannot rule out so easily the possibility of finding a final theory: he made a dogmatic and unwarranted statement.

8. M. Bunge, *Causality: The Place of the Causal Principle in Modern Science* (Cambridge: Harvard University Press, 1959).

9. See especially W. C. Salmon, *Scientific Explanation and the Causal Structure of the World* (Princeton: Princeton University Press, 1984).

10. Hacking, *Representing and Intervening*, 167.

11. Ibid., 180–81.

12. W. V. O. Quine, "Existence," in *Physic, Logic and History*, ed. W. Y. Yourgrau and A. D. Breck (New York and London: Plenum Press, 1970), 94; italics added.

13. Ibid., 93–94.

14. W. V. O. Quine, "On What There Is," in *From a Logical Point of View*, 16–17.

15. As previously remarked in note 1 of this chapter, only a few authors, such as A. Orenstein and N. Rescher, have openly questioned

the Fregean paradigm. But see also the volume by M. K. Munitz, *Existence and Logic* (New York: New York University Press, 1974), for an extensive criticism of Frege's theses.

16. Hacking, *Representing and Intervening*, 101–2.

17. It should be noted, though, that Putnam's position evolves quickly, and his latest works pay much attention to the theses of Peirce and other representatives of American pragmatism.

18. This is, for instance, the opinion of E. Agazzi. See his article, "La questione del realismo scientifico," in *Saggi in onore di Ludovico Geymonat*, ed. C. Mangione (Milan: Garzanti, 1985), 171–92.

19. This has been done, for example, by Peter Smith in his book *Realism and the Progress of Science* (Cambridge: Cambridge University Press, 1981).

20. See chapter 1, section 4.

21. W. V. O. Quine, "Identity and Hypostasis," in *From a Logical Point of View*, 78–79.

22. As Steven Weinberg points out, these disputes are not popular among scientists: they usually share a basic realism. I would like to add that this happens because, acting that way, they get things *done*.

23. See chapter 1, section 5.

24. See chapter 1, section 4.

25. D. Pears, *The False Prison* (Oxford: Oxford University Press, 1988).

26. Wittgenstein, *Tractatus*, 3.221–23.

27. It might be argued in this regard that Wittgenstein was influenced by Arthur Schopenhauer, whose works he read carefully.

CHAPTER 3

1. See chapter 2, sections 1 and 2.

2. I. Kant, *Critique of Pure Reason* (New York: St. Martin's Press, 1965), A 493-B 521, pp. 440–41.

3. Hacking, *Representing and Intervening*, 21–24.

4. Putnam, *Il pragmatismo*, 67–69.

5. Putnam, along with many others, often invokes the case of quantum theory for supporting anti-realist theses. It should be observed, however, that this theory may also have a realistic interpretation; see for instance H. Krips, *The Metaphysics of Quantum Theory*, 2d ed. (Oxford: Clarendon Press, 1990).

6. Some colleagues, among them Wesley Salmon and Malcom Macmillan, objected to me that this is not a sound example because we have good reasons to believe that Columbus was not the first one to reach the American continent. Most likely the Vikings made the trip several centu-

ries before him, not to mention the popular tales concerning the Irish monk St. Brendan. It should be clear, however, that this is not relevant to my argument. Even leaving out the fact that the value of a geographic discovery can be judged only when the discoverer informs the world about what he found out, I chose Columbus because he is normally identified as the one who did the job. But I also know that, after all, the native Americans were there many thousand years prior to Columbus. My argument is philosophical, not historical: nothing would change by replacing Columbus with Erik the Red or St. Brendan. It seems to me that this objection is too much influenced by polemics that took place, both in Europe and the United States, during the celebrations of the five hundredth anniversary of the discovery of America. It is easily understood, however, that those polemics have nothing to do with philosophical argumentations.

7. I wish to thank Nicholas Rescher for raising this objection and the following one during a lecture I delivered at the Center for Philosophy of Science of the University of Pittsburgh.

8. Weinberg, *Final Theory*, 173.

9. N. Rescher, *Scientific Realism: A Critical Reappraisal* (Dordrecht and Boston: Reidel, 1987), 6.

10. Just recall what I said before, in note 7 to chapter 2, about Hilary Putnam's claim that we will never be able to find a final theory in the sense attributed to this expression by the physicist Steven Weinberg.

11. Rescher, *Scientific Realism*, 30–31.

12. See for instance the interesting article by M. Bartusiak, "Loops of Space," *Discover* 14, no. 3 (1993): 60–68.

13. Rescher, *Scientific Realism*, xii–xiii.

14. Even Richard Rorty often speaks of the analytic "reification of language." But, as we shall see later on, the conclusions he draws from this insight are quite different from ours.

15. Weinberg, *Final Theory*, chapter 3.

16. See note 2 of this chapter.

17. Ibid.

18. Recall what we said in the second chapter about the Kantian stance endorsed by Quine.

19. M. Marsonet, "Is Philosophy of Language Really Important for the Foundation of Scientific Realism?" *American Philosophical Quarterly* 30, no. 4 (1993): 228–301.

20. B. Russell, *A History of Western Philosophy* (New York: Simon and Schuster, 1945), 707–8; italics added.

21. See for instance the article by K. Baldner, "Is Transcendental Idealism Coherent?" *Synthese* 85 (1990): 1–23.

22. Recall the aforementioned astonishment shown by Nobel prize-winning physicist Steven Weinberg when he reports that most contempo-

rary philosophers ignore Einstein's theories and prefer to stick to Kant's conception of space and time (see note 8 of this chapter).

23. H. Reichenbach, *The Rise of Scientific Philosophy* (New York: Dover, 1966), 48–49.

24. H. Skolimowksi, *Living Philosophy: Eco-Philosophy as a Tree of Life* (New York and London: Penguin Books, 1992), 87–89.

25. Rescher, *Scientific Realism.*

26. Since I do not want to be accused of forgetting Great Britain and the United States, let us recall that even in these two countries there have been important idealist thinkers. Bradley and McTaggart are two outstanding examples, while both Bertrand Russell and John Dewey assumed idealist positions in the first stages of their philosophical parabolae. But, mainly for cultural reasons, the British and American forms of idealism never had an influence comparable to the one idealism enjoyed—and still enjoys—in many countries of continental Europe.

27. J. O. Urmson and J. Ree, eds., *The Concise Encylopedia of Western Philosophy and Philosophers*, rev. ed. (London: Unwin Hyman, 1989), 146–49.

28. N. Rescher, *Conceptual Idealism* (Oxford: Blackwell, 1973), 161–67.

29. Rescher, *Scientific Realism*, 147–48.

30. Ibid., 151.

31. This and the following quotations are drawn from Rescher's *Scientific Realism*, 147.

CHAPTER 4

1. Rescher, *Conceptual Idealism* and *Scientific Realism.*

2. N. Rescher, "Conceptual Idealism Revisited," *Review of Metaphysics* 44, no. 3 (1991): 495–523.

3. See my paper "Is Philosophy of Language Really Important for the Foundation of Scientific Realism"; and chapter 2, section 5, of this book.

4. Rescher, "Conceptual Idealism Revisited," 495.

5. See for instance the definition of idealism contained in J. H. Fetzer and R. F. Almeder, *Glossary of Epistemology/Philosophy of Science* (New York: Paragon House, 1993), 65.

6. See his volume *Scientific Realism*, chapters 2, 3, and passim.

7. Rescher, "Conceptual Idealism Revisited," 495.

8. See for instance the article by D. Sobel, "Man Stops Universe, Maybe," *Discover* 14, no. 4 (1993): 20–21. In it, a story is narrated about William Tifft, an astronomer at the University of Arizona, whose empirical data seem to deny the expansion of the universe. If so, the Big

Bang theory collapses and the whole picture of the universe endorsed by today's cosmology would be, more or less, untenable.

9. Weinberg, *Final Theory*.

10. Rescher, *Scientific Realism*, 38–39.

11. Ibid., 41.

12. See the book by K. R. Popper and K. Lorenz, *Die Zukunft ist offen* (Munich: R. Piper, 1985).

13. See especially P. Teilhard de Chardin, *Le phénomène humain* (Paris: Seuil, 1955) (English translation: *The Phenomenon of Man* [New York, Harper and Brothers, 1959]).

14. Rescher, "Conceptual Idealism Revisited," 496–97.

15. Note that there is continuity even between us and such typical fictional objects as "Pegasus." To deny that Pegasus exists despite the fact that (a) mankind created this ficitional character and (b) it can be found in our mythology texts is, in my view, unforgiveable.

16. See chapter 2, section 1.

17. See again my paper, "Is Philosophy of Language Really Important for the Foundation of Scientific Realism?" especially sections 6 and 7.

18. Rescher, "Conceptual Idealism Revisited," 499–500.

19. See especially sections 2–5.

20. Fetzer and Almeder, *Glossary*, 65.

21. Rescher, "Conceptual Idealism Revisited," 512.

22. Ibid., 518.

23. Ibid., 516–17; italics added.

24. Mankind was not born with thought and language, so to speak, "incorporated." In claiming that, we carry on a "reification" of language and thought. To adopt a priori concepts of language and thought means to *hypostatize* them: they become something which is out of time, out of space, out of physical reality, and out of everything.

25. Ibid., 521–23.

26. The aforementioned remarks are contained in a letter that Rescher addressed to me on 12 April 1993.

CHAPTER 5

1. Rorty, "Wittgenstein," 337.

2. As a matter of fact, Rorty in his paper addresses Heidegger's thought, too, but I will not take Heidegger into account in my work.

3. This is Rorty's own formulation of the aforementioned referential problem: "If we claim that no entity is available which remains unrelated by a form of relationship which cannot hold between unaided type

B entitities, then we have problems about the availability of the type A entities we postulate to lend the necessary aid" (Ibid., 342).

4. See especially Wittgenstein's *On Certainty.*

5. Rorty, "Wittgenstein," 342.

6. Ibid., 343.

7. Hacking, *Representing and Intervening*, 63.

8. Rorty, "Wittgenstein," 354.

9. Fetzer and Almeder, *Glossary*, 96–97.

10. Wittgenstein, *On Certainty*, 96–99, p. 15e.

11. See for instance R. Monk, *Ludwig Wittgenstein: The Duty of Genius* (London: Jonathan Cape, 1990), 569–75.

12. Ibid., 578.

13. Wittgenstein, *Tractatus*, 2.0211–0212.

14. Monk, *Wittgenstein*, 563.

15. Rorty, "Wittgenstein," 345–46.

16. Wittgenstein, *On Certainty*, 310, p. 40e.

17. R. Rorty, *Ibid.*, 340.

18. *Astronomy* 21, no. 3 (March 1993): 13.

19. See chapter 1, section 5.

20. See chapter 2, section 1.

21. Weinberg, *Final Theory*, 5–6; italics added.

22. F. Nietzsche, *Thus Spoke Zarathustra*, English trans., 2d ed. (London: Penguin Books, 1969), 39.

23. Weinberg, *Final Theory*, 9.

24. As I remarked before, a typical case is Hilary Putnam.

25. Ibid., 27–29.

26. Ibid., 14.

27. P. F. Strawson, *Analysis and Metaphysics* (New York and Oxford: Oxford University Press, 1992), 14–15.

28. Weinberg, *Final Theory*, 16.

29. See W. C. Salmon, *Scientific Explanation.*

30. Weinberg, *Final Theory*, 25.

31. Ibid., 188.

32. Ibid., 190.

33. Ibid., 47–48.

34. Ibid., 93.

35. W. Balzer, D. A. Pearce, and H.-J. Schmidt, eds., *Reduction in Science: Structure, Examples, Philosophical Problems* (Dordrecht, Reidel, 1984).

36. Weinberg, *Final Theory*, 168.

37. Ibid., 61

38. See especially chapters 1 and 2.

39. W. Sellars, "Language as Thought and as Communication," *Philosophy and Phenomenological Research* 29 (1969): 506–27. See also

W. Sellars, "Empiricism and the Philosophy of Mind," in *Science, Perception and Reality* (London: Routledge and Kegan Paul, 1963); and W. Sellars and R. Chisholm, "Intentionality and the Mental," in *Minnesota Studies in the Philosophy of Science*, vol. 2, ed. H. Feigl and M. Scriven (Minneapolis, 1957), 507–39.

40. P. Kitcher, "The Naturalists Return," *Philosophical Review* 101, no. 1 (1992): 53–55.

BIBLIOGRAPHY

Adorno, T. W., K. R. Popper, et al. *Der Postivismusstreit in der deutschen Soziologie*. Berlin: H. Luchterhand Verlag, 1969.

Agazzi, E. "La questione del realismo scientifico." In *Saggi in onore di Ludivico Geymonat*, edited by C. Mangione, 171–92. Milan: Garzanti, 1985.

———. "Subjectivity and Ontological Committment in the Empirical Sciences." In *Historical and Philosophical Dimensions of Logic, Methodology and Philosophy of Science*, edited by A. Butts and J. Hintikka, 159–71. Dordrecht: Reidel, 1977.

Ajdukiewicz, K. *Problems and Theories of Philosophy*. Cambridge: Cambridge University Press, 1973.

———. *The Scientific World-Perspective and Other Essays: 1931–1963*. Edited by J. Giedymin. Dordrecht: Reidel, 1978.

Almog, J. "Logic and the World." In *Themes from Kaplan*, edited by J. Almog, J. Perry, and H. Wettstein, 43–65. New York and Oxford: Oxford University Press, 1989.

Anscombe, G. E. M. "The Question of Linguistic Idealism." In *The Philosophy of Wittgenstein*, vol. 8, *Knowing, Naming, Certainty and Idealism*, edited by J. V. Canfield, 188–215. New York and London: Garland, 1986.

Aronson, J. *A Realist Philosophy of Science*. New York: St. Martin's Press, 1984.

Ayer, A. J. *The Central Questions of Philosophy*. London: Weidenfeld and Nicholson, 1973.

———. *Language, Truth and Logic*. London: Macmillan, 1936. Reprint, London: Penguin Books, 1990.

———. *The Meaning of Life and Other Essays*. London: Weidenfeld and Nicholson, 1990.

———. *Philosophy in the Twentieth Century*. London: Weidenfeld and Nicholson, 1982.

———, ed. *Logical Positivism*. New York: The Free Press, 1959.

Baldner, K. "Is Transcendental Idealism Coherent?" *Synthese* 85 (1990): 1–23.

Balzer, W., D. A. Pearce, and H.-J. Schmidt. *Reduction in Science: Structure, Examples, Philosophical Problems*. Dordrecht: Reidel, 1984.

Bartusiak, M. "Loops of Space." *Discover* 14, no. 3 (1993): 60–68.

Bergmann, G. *The Metaphysics of Logical Positivism*. 2d ed. Madison: University of Wisconsin Press, 1967.

Bloor, D. *Wittgenstein: A Social Theory of Knowledge*. London: Macmillan, 1983.

Braithwaite, R. B. *Scientific Explanation: A Study of the Function of Theory, Probability and Law in Science*. Cambridge: Cambridge University Press, 1953.

Brandom, R. "Reference Explained Away." *Journal of Philosophy* 81 (1984): 469–92.

———. "Truth and Assertibility." *Journal of Philosophy* 73 (1976): 137–49.

Bunge, M. *Causality: The Place of the Causal Principle in Modern Science*. Cambridge: Harvard University Press, 1959.

Burkhardt, H., and B. Smith, eds. *Handbook of Metaphysics and Ontology*. Philadelphia, Munich, and Vienna: Philosophia Verlag, 1991.

Bynum, W. F., E. J. Browne, and R. Porter, eds. *Dictionary of the History of Science*. 2d. ed. Princeton: Princeton University Press, 1984.

Callebaut, W., ed. *Taking the Naturalistic Turn, or How Real Philosophy of Science is Done*. Chicago and London: University of Chicago Press, 1993.

Carnap, R. "The Elimination of Metaphysics through Logical Analysis of Language." In *Logical Postivism*, edited by A. J. Ayer, 60–81. New York: The Free Press, 1959.

———. *The Logical Syntax of Language*. London: Routledge and Kegan Paul, 1937.

———. *Meaning and Necessity*. Chicago: University of Chicago Press, 1947.

Cartwright, N. *How the Laws of Physics Lie*. Oxford: Clarendon Press, 1983.

Churchland, P. M. *Scientific Realism and the Plasticity of Mind*. 4th ed. Cambridge: Cambridge University Press, 1990.

Coffa, J. A. *The Semantic Tradition from Kant to Carnap, to the Vienna Station*. Cambridge: Cambridge University Press, 1991.

Davidson, D. "A Coherence Theory of Truth and Knowledge." In *Truth and Interpretation: Perspectives on the Philosophy of Donald Davidson*, edited by E. LePore, 307–19. Oxford: Blackwell, 1986.

———. "Empirical Content." In *Truth and Interpretation. Perspectives on the Philosophy of Donald Davidson*, edited by E. LePore, 320–32. Oxford: Blackwell, 1986.

———. "A Nice Derangement of Epitaphs." In *Truth and Interpretation. Perspectives on the Philosophy of Donald Davidson*, edited by E. LePore, 433–46. Oxford: Blackwell, 1986.

———. "On the Very Idea of a Conceptual Scheme." *Proceedings of the American Philosophical Association* 47 (1974): 5–20.

Davidson, D., and J. Hintikka, eds. *Words and Objections: Essays on the Work of W. V. Quine*. Boston and Dordrecht: Reidel, 1969.

Delaney, C. F., M. J. Loux, G. Gutting, and W. D. Solomon, eds. *The Synoptic Vision: Essays on the Philosophy of Wilfrid Sellars*. Notre Dame and London: University of Notre Dame Press, 1977.

Dewey, J. *Logic, the Theory of Inquiry*. New York: H. Holt and Co., 1938.

Dummett, M. 1973. *Frege: Philosophy of Language*. London: Duckworth.

———. "A Nice Derangement of Epitaphs: Some Comments on Davidson and Hacking." In *Truth and Interpretation. Perspectives on the Philosophy of Donald Davidson*, edited by E. LePore, 459–76. Oxford: Blackwell, 1986.

Dunn, J. M., and N. D. Belnap. "The Substitution Interpretation of the Quantifiers." *Nous* 2 (1968): 177–185.

Earman, J. "Against Indeterminacy." *Journal of Philosophy* 74 (1977): 535–38.

Evans, G., and J. McDowell, eds. *Truth and Meaning*. Oxford: Clarendon Press, 1976.

Faris, J. A. *Quantification Theory*. London: Routledge and Kegan Paul, 1964.

Fetzer, J. H., and R. F. Almeder. *Glossary of Epistemology/Philosophy of Science*. New York: Paragon House, 1993.

Feyerabend, P. K. *Against Method*. London: New Left Books, 1975.

———. *Dialogo sul metodo*. Italian trans. Rome and Bari: Laterza, 1989.

———. *Three Dialogues on Knowledge*. Oxford: Blackwell, 1991.

Flower, E., and M. G. Murphey. *A History of Philosophy in America*. New York: Capricorn Books, 1977.

Forbes, G. *The Metaphysics of Modality*. Oxford: Clarendon Press, 1985.

Glymour, C. "Realism and the Nature of Theories." In *Introduction to the Philosophy of Science*, edited by M. H. Salmon et al., 104–31. Englewood Cliffs, NJ: Prentice Hall, 1992.

Gochet, P. *Ascent to Truth: An Essay on Quine's Philosophy*. Munich and Vienna: Philosophia Verlag, 1986.

Goodman, N. *The Structure of Appearance*. Cambridge: Harvard University Press, 1951.

Gosselin, M. *Nominalism and Contemporary Nominalism: Ontological and Epistemological Implications of the Work of W. V. O. Quine and of N. Goodman*. Boston and Dordrecht: Kluwer, 1990.

Gower, B., ed. *Logical Positivism in Perspective*. London and Sydney: Croom Helm, 1987.

Grünbaum, A. *Philosophical Problems of Space and Time*. 2d ed. Boston and Dordrecht: Reidel, 1974.

Haack, S. *Philosophy of Logics*. Cambridge: Cambridge University Press, 1978.

———. "Realism." *Synthese* 73 (1987): 275–99.

Hacking, I. *Representing and Intervening: Introductory Topics in the Philosophy of Natural Science*. 8th ed. Cambridge: Cambridge University Press, 1993.

———. *Why Does Language Matter to Philosophy?* Cambridge: Cambridge University Press, 1975.

Hahn, H., O. Neurath., and R. Carnap. *Wissenschaftliche Weltauffassung*. Vienna: Der Wiener Kreis, 1929.

Hahn, L. E., and P. A. Schilpp, eds. *The Philosophy of W. V. Quine*. La Salle, IL: Open Court, 1986.

Harre', R. *Varieties of Realism: A Rationale for the Natural Sciences*. Oxford: Blackwell, 1986.

Hawking, S. *A Brief History of Time: From the Big Bang to Black Holes*. New York: Bantam Books, 1988.

Hempel, C. G. *Aspects of Scientific Explanation and Other Essays in the Philosophy of Science*. New York: The Free Press, 1965.

Kant, I. *Critique of Pure Reason*. New York: St. Martin's Press, 1965.

Kearns, K. T. "The Logical Concept of Existence." *Notre Dame Journal of Formal Logic* 9, no. 4 (1968): 313–24.

Kenny, A. J. P. *Wittgenstein*. London: Penguin Books, 1973.

Kitcher, P. "The Naturalists Return." *Philosophical Review* 101, no. 1 (1992): 53–114.

Kitcher, P., and W. C. Salmon, eds. *Scientific Explanation*. Minnesota Studies in the Philosophy of Science, vol. 13. Minneapolis: University of Minnesota Press, 1989.

Knuuttila, S., and J. Hintikka, eds. *The Logic of Being*. Dordrecht: Reidel, 1986.

Kolakowski, L. *Positivist Philosophy: From Hume to the Vienna Circle*. London: Penguin Books, 1972.

Krips, H. *The Metaphysics of Quantum Theory*. 2d ed. Oxford: Clarendon Press, 1990.

Kuhn, T. S. *The Structure of Scientific Revolutions*. 2d ed. Chicago: University of Chicago Press, 1970.

Küng, G. "Ajdukiewicz's Contribution to the Realism/Idealism Debate." In *The Vienna Circle and Lvov-Warsaw School*, edited by K. Szaniawski, 67–85. Dordrecht: Kluwer, 1989.

———. *Ontology and the Logistic Analysis of Language*. Dordrecht: Reidel, 1967.

Lakatos, I., and A. Musgrave, eds. *Criticism and the Growth of Knowledge*. Cambridge: Cambridge University Press, 1970.

Lejewski, C. "Logic and Existence." *British Journal for the Philosophy of Science* 5 (1954): 104–19.

Lepore, E., ed. *Truth and Interpretation: Perspectives on the Philosophy of Donald Davidson.* Oxford: Blackwell, 1986.

Lessnoff, M. H. *The Structure of Social Science: A Philosophical Introduction.* London: Allen and Unwin, 1974.

Linksy, L., ed. *Semantics and the Philosophy of Language.* Urbana: University of Illinois Press, 1952.

Loux, M., ed. *The Possible and the Actual: Readings in the Metaphysics of Modality.* Ithaca and London: Cornell University Press, 1979.

———. *Universals and Particulars: Readings in Ontology.* Notre Dame and London: University of Notre Dame Press, 1976.

Malcom, N. "Wittgenstein and Idealism." In *The Philosophy of Wittgenstein*, vol. 8, *Knowing, Naming, Certainty and Idealism*, edited by J. V. Canfield, 339–57. New York and London: Garland, 1986.

Marsonet, M. "Is Philosophy of Language Really Important for the Foundation of Scientific Realism?" *American Philosophical Quarterly* 30, no. 4 (1993): 283–301.

———. *Linguaggio e conoscenza: Saggio su K. Ajdukiewicz.* Milan: Angeli, 1986.

———. "Realism and Anti-Realism: An Old/New Debate." *Idealistic Studies* 23 (1993): 123–37.

———. "Réflexions sur le conventionnalisme radical de K. Ajdukiewicz." In *Aspects et problèmes du conventionnalisme*, edited by G. Sommaruga-Rosolemos, 113–24. Fribourg: Editions universitaires de Fribourg, 1992.

Massey, G. J. "Indeterminacy, Inscrutability, and Ontological Reference." In *Studies in Ontology*, American Philosophical Quarterly Monograph Series, no. 12, pp. 43–55. Oxford: Blackwell, 1978.

McGuinness, B. *Wittgenstein: A Life. The Young Ludwig, 1889–1921.* Berkeley and Los Angeles: University of California Press, 1988.

Mises, R. von. *Positivism.* English trans. New York: Dover, 1968.

Monk, R. *Ludwig Wittgenstein: The Duty of Genius.* London: Jonathan Cape, 1990.

Mook, D. E., and T. Vargish. *Inside Relativity.* Princeton: Princeton University Press, 1987.

Moore, G. E. "Is Existence a Predicate?" In *Philosophical Papers*, 115–25. London: Allen and Unwin, 1959.

———. "Proof of an External World." In *Philosophical Papers*, 126–48. London: Allen and Unwin, 1959.

Munitz, M. K. *Existence and Logic.* New York: New York University Press, 1974.

———. *The Question of Reality.* Princeton: Princeton University Press, 1990.

Murphy, J. P. *Pragmatism: From Peirce to Davidson.* Boulder, San Francisco, and Oxford: Westview Press, 1990.

Nagel, E. *The Structure of Science: Problems in the Logic of Scientific Explanation.* New York: Harcourt, Brace and World, 1961.

Neurath, O. *Philosophical Papers, 1913–1946.* Dordrecht: Reidel, 1983.

Nietzsche, F. *Thus Spoke Zarathustra.* 2d ed. London: Penguin Books, 1969.

Norton, J. D. "Philosophy of Space and Time." In *Introduction to the Philosophy of Science*, edited by M. H. Salmon et al., 179–231. Englewood Cliffs, NJ: Prentice-Hall, 1992.

Oldroyd, D. *The Arch of Knowledge: An Introductory Study of the History, Philosophy, and Methodology of Science.* New York: Methuen, 1986.

Orenstein, A. *Existence and the Particular Quantifier.* Philadelphia: Temple University Press, 1978.

Passmore, J. *A Hundred Years of Philosophy.* 2d ed. London: Penguin Books, 1966.

Pears, D. 1988. *The False Prison.* Oxford: Oxford University Press, 1988.

Phillipps, S. "Counting to the Edge of the Universe." *Astronomy* 21, no. 4 (1993): 38–43.

Pitt, J. C., ed. *The Philosophy of Wilfrid Sellars: Queries and Extensions.* Boston and Dordrecht: Reidel, 1978.

Plantinga, A. *The Nature of Necessity.* Oxford: Clarendon Press, 1974.

Popper, K. R. *Conjectures and Refutations.* London: Routledge and Kegan Paul, 1963.

———. *The Logic of Scientific Discovery.* New York: Basic Books, 1959. Reprint, rev. ed., London: Hutchinson, 1968.

———. *Objective Knowledge.* Oxford: Oxford University Press, 1972.

———. "A Realist View of Logic, Physics, and History." In *Physics, Logic, and History*, edited by W. Y. Yourgrau and A. D. Breck, 1–37. New York: Plenum Press, 1970.

———. *Unended Quest: An Intellectual Autobiography.* London: Fontana, 1976.

Popper, K. R., and K. Lorenz. *Die Zukunft ist offen.* Munich: R. Piper, 1985.

Przelecki, M. "The Approach to Metaphysics in the Lvov-Warsaw School." In *The Vienna Circle and the Lvov-Warsaw School*, edited by K. Szaniawski, 55–66. Dordrecht: Kluwer, 1989.

Putnam, H. *Il pragmatismo: una questione aperta.* Rome and Bari: Laterza, 1992.

———. *The Many Faces of Realism.* La Salle, IL: Open Court, 1987.

———. *Meaning and the Moral Sciences.* London: Routledge and Kegan Paul, 1978.

———. *Realism with a Human Face*. Cambridge: Harvard University Press, 1990.

———. *Reason, Truth and History*. Cambridge: Cambridge University Press, 1981.

Quine, W. V. O. "Existence." In *Physics, Logic and History*, edited by W. Y. Yourgrau and A. D. Breck, 89–103. New York and London: Plenum Press, 1970.

———. *From a Logical Point of View*. New York: Harper and Row, 1953.

———. *Mathematical Logic*. Cambridge: Harvard University Press, 1976.

———. *Methods of Logic*. Rev. ed. New York: Holt, Rinhart and Winston, 1959.

———. *Ontological Relativity and Other Essays*. New York: Columbia University Press, 1969.

———. *Philosophy of Logic*. Englewood Cliffs, NJ: Prentice-Hall, 1970.

———. *Pursuit of Truth*. 2d ed. Cambridge University Press, 1992.

———. *Quiddities*. Cambridge: Harvard University Press, 1987.

———. *The Roots of Reference*. La Salle, IL: Open Court, 1974.

———. *The Ways of Paradox and Other Essays*. New York: Random House, 1966.

———. *Word and Object*. Cambridge: MIT Press, 1960.

Ray, C. *Time, Space and Philosophy*. New York and London: Routledge, 1991.

Reichenbach, H. *The Philosophy of Space and Time*. New York: Dover, 1957.

———. *The Rise of Scientific Philosophy*. New York: Dover, 1966.

———. *Selected Writings: 1909–1953*. Dordrecht: Reidel, 1978.

Rescher, N. *The Coherence Theory of Truth*. Oxford: Clarendon Press, 1973.

———. *Conceptual Idealism*. Oxford: Blackwell, 1973.

———. "Conceptual Idealism Revisited." *Review of Metaphysics* 44, no. 3 (1991): 495–523.

———. "Conceptual Schemes." In *Studies in Epistemology*, Midwest Studies in Philosophy, no. 5, edited by P. A. French, E. Uehling, and H. K. Wettstein, 323–45. Minneapolis: University of Minnesota Press, 1979.

———. "The Equivocality of Existence." In *Studies in Ontology*, American Philosophical Quarterly Monograph Series, no. 12, pp. 57–66. Oxford: Blackwell, 1978.

———. *Methodological Pragmatism*. Oxford: Blackwell, 1977.

———. *Scientific Progress*. Oxford: Blackwell, 1978.

———. *Scientific Realism: A Critical Reappraisal*. Boston and Dordrecht: Reidel, 1987.

———. *A System of Pragmatic Idealism.* Vol 1, *Human Knowledge in Idealistic Perspective.* Princeton: Princeton University Press, 1992.

Romanos, G. D. *Quine and Analytic Philosophy: The Language of Language.* 2d ed. Cambridge: MIT Press, 1984.

Rorty, R. *Consequences of Pragmatism.* Minneapolis: University of Minnesota Press, 1982.

———. *Philosophical Papers.* Vol. 1, *Objectivity, Relativism, and Truth.* Cambridge: Cambridge University Press, 1991.

———. *Philosophy and the Mirror of Nature.* Princeton: Princeton University Press, 1979.

———. "Pragmatism, Davidson and Truth." In *Truth and Interpretation: Perspectives on the Philosophy of Donald Davidson*, edited by E. Lepore, 333–55. Oxford: Blackwell, 1986.

———. "Wittgenstein, Heidegger, and the Reification of Language." In *The Cambridge Companion to Heidegger*, edited by C. B. Guignon, 337–57. Cambridge: Cambridge University Press, 1993.

———, ed. *The Linguistic Turn.* Chicago: University of Chicago Press, 1967.

Russell, B. *History of Western Philosophy.* New York: Simon and Schuster, 1945.

———. "Logical Atomism." In *Logical Positivism*, edited by A. J. Ayer, 31–50. New York: The Free Press, 1959.

———. *Our Knowledge of the External World.* London: Allen and Unwin, 1914.

———. *The Problems of Philosophy.* New York: H. Holt and Company, 1912.

———. *The Scientific Outlook.* London: Allen and Unwin, 1931.

Salmon, M. H., et al., eds. *Introduction to the Philosophy of Science.* Englewood Cliffs, NJ: Prentice-Hall, 1992.

Salmon, W. C. *The Foundations of Scientific Interference.* 5th ed. Pittsburgh: University of Pittsburgh Press, 1979.

———. *Four Decades of Scientific Explanation.* Minneapolis: University of Minnesota Press, 1990.

———. *Scientific Explanation and the Causal Structure of the World.* Princeton: Princeton University Press, 1984.

Scheffler, I. *The Anatomy of Inquiry.* New York: Alfred Knopf, 1963.

Sellars, W. "Empiricism and the Philosophy of Mind." In *Minnesota Studies in the Philosophy of Science*, vol. 1, edited by H. Feigl and M. Scriven. Minneapolis: University of Minnesota Press, 1956. Reprinted as chapter 5 in W. Sellars, *Science, Perception and Reality*, below.

———. "Language as Thought and as Communication." *Philosophy and Phenomenological Research* 29 (1969): 506–527, 1969.

———. *Naturalism and Ontology.* Atascadero, CA: Ridgeview Publishing Co., 1979.

———. *Science and Metaphysics: Variations on Kantian Themes.* London: Routledge and Kegan Paul, 1968.

———. *Science, Perception and Reality.* London: Routledge and Kegan Paul, 1963.

Sellars, W., and R. Chisholm. "Intentionality and the Mental." In *Minnesota Studies in the Philosophy of Science*, vol. 2, edited by H. Feigl and M. Scriven, 507–39. Minneapolis: University of Minnesota Press, 1957.

Sklar, L. *Space, Time, and Spacetime.* 3d ed. Berkeley and Los Angeles: University of California Press, 1977.

Skolimowski, H. *Living Philosophy: Eco-Philosophy as a Tree of Life.* New York and London: Penguin Books, 1992.

———. *Polish Analytical Philosophy.* London: Routledge and Kegan Paul, 1967.

———. "Quine, Adjukiewicz, and the Predicament of Twentieth-Century Philosophy." In *The Philosophy of* W. V. *Quine*, edited by L. E. Hahn and P. A. Schilpp, 463–93. La Salle, IL: Open Court, 1986.

Smith, P. *Realism and the Progress of Science.* Cambridge: Cambridge University Press, 1981.

Sobel, D. "Man Stops Universe, Maybe." *Discover* 14, no. 4 (1993): 20–21.

Strawson, P. F. *Analysis and Metaphysics.* New York and Oxford: Oxford University Press, 1992.

———. *Introduction to Logical Theory.* London: Methuen, 1952.

———. "Is Existence Never a Predicate?" In *Freedom and Resentment and Other Essays*, 189–97. London: Methuen, 1974.

———, ed. *Philosophical Logic.* Oxford: Clarendon Press, 1967.

Szaniawski, K., ed. *The Vienna Circle and the Lvov-Warsaw School.* Dordrecht: Kluwer, 1989.

Taylor, R. *Metaphysics.* 2d ed. Englewood Cliffs, NJ: Prentice-Hall, 1974.

Teilhard de Chardin, P. *The Phenomenon of Man.* New York: Harper and Brothers, 1959.

Thomas, D. *Naturalism and Social Science: A Post-Empiricist Philosophy of Social Science.* Cambridge: Cambridge University Press, 1979.

Urmson, J. O. *Philosophical Analysis.* Oxford: Clarendon Press, 1960.

Urmson, J. O., and J. Ree, eds. *The Concise Encyclopedia of Western Philosophy and Philosophers.* Rev. ed. London: Unwin Hyman, 1989.

Van Fraasen, B. C. *The Scientific Image.* 5th ed. Oxford: Clarendon Press, 1990.

Weinberg, J. R. *An Examination of Logical Positivism.* London: Paul, Trench and Trubner, 1936.

Weinberg, S. *Dreams of a Final Theory.* New York: Pantheon Books, 1992.

Williams, B. "Wittgenstein and Idealism." In *The Philosophy of Wittgenstein*, vol. 8, *Knowing, Certainty and Idealism*, edited by J. V. Canfield, 318–37. New York and London: Garland, 1986.

Williams, C. J. F. *What is Existence?* Oxford: Clarendon Press, 1981.

Winch, P. *The Idea of a Social Science and Its Relations to Philosophy*. London: Routledge and Kegan Paul, 1958.

Wittgenstein, L. *On Ceriainty*. English trans. Oxford: Blackwell, 1969.

———. *Philosophical Investigations*. New York and London: Macmillan, 1958.

———. *Tractatus Logico-Philosophicus*. Edited by A. J. Ayer. London: Routledge and Kegan Paul, 1961.

INDEX OF NAMES

INDEX OF SUBJECTS